# SUBMITTING A WINNING BID

## Guide to Making a Construction Bidding with Examples

# COPYRIGHT

While we have taken every precaution in preparing this book, the Publisher assumes no responsibility for errors or omissions, or damages resulting from the use of the information contained.

## SUBMITTING A WINNING BID

Copyright © 2021 Gustavo Cinca

Written by Gustavo Cinca

- Copyright ........................................................................... 3
- PREFACE ............................................................................ 5
- Review of Tender Documentation ............................................ 9
- Engineering Cost Evaluation ................................................. 15
- Inventory of Materials .......................................................... 18
- Estimating Direct Man-hours ................................................ 24
- Assess the Kind of Equipment ............................................... 31
- Estimation of Indirect Labor ................................................. 45
- Overhead Costs .................................................................... 53
- Calculation of the Cost of Insurance ..................................... 59
- Analysis of finance costs ...................................................... 81
- Contingency Reserve ............................................................ 84
- Profit Margin ....................................................................... 89
- Cost of Taxes ....................................................................... 98
- Example of a Direct Estimate of Person-hours ..................... 101
- Example of Calculating the Cost of Equipment per Hour ... 104
- Financial Cost Example ....................................................... 112

# PREFACE

For a construction and industrial assembly business to be financially viable, it must have positive economic outcomes in the work or services it provides.

When the assigned project originates from a bid with errors, the project will not be profitable.

To launch a lucrative business, the bidder must submit a properly evaluated bid in each bidding process.

In this book "Submitting a Winning Bid" we are going to define the steps to follow to prepare a reliable bid.

This manuscript is of particular interest to owners, shareholders and tender coordinators of construction and industrial assembly, and in general to all members of an organization who perform tasks related to the formulation of tenders or price competitions.

Quotation with certainty is a decisive factor in performing a lucrative contract.

Bidding at prices far from the market average undermines the commercial relationship between the proposer and the customer.

When the bidder submits a budget that is too low, it will inevitably face negative financial outcomes as expenditure will exceed revenues.

On the other hand, if the supply is of remarkably high value in relation to the proposals of the competition, it will certainly be out of price competition.

This last situation, although less expensive than the previous one, affects their finances because of the increase in general spending.

The conclusion is that in quoting, we must carefully study each stage to present a reliable bid.

This publication provides the reader with a complete and useful guide to assist him in his budgeting.

The development of every economic proposal is undoubtedly a decisive task in the management of a construction and assembly company.

In this manuscript, the aspects to be considered at each stage of developing proposals to achieve a reasonable budget are detailed.

Examples of applications can be found at the end of the book.

## Budgeting requires

The individual or group responsible for preparing the submission should be diligent and carefully consider the proposal to eliminate errors.

Review and verify conformance to specification requirements.

To have own experience in the construction and assembly of the work to be estimated, or in its absence, the ability to use emotional intelligence and reconcile criteria with other colleagues or specialists to increase the knowledge of what is estimated.

Know the resources available and the capacity of your organization.

Know how to apply worker performance charts, what equipment to use in the project, etc.

This manual provides technical support to budgeters and is based on the author's exceptional experience.

The author of this book has worked for a large part of his professional life as a manager and director of on-site work at various chemical processing plants, refineries, gas pipelines, gas compressors and thermal power stations in the country and abroad, finally, to create and to chair a society of construction and assembly.

Throughout his career, the author has prepared and reviewed hundreds of estimates of bids for the refurbishment of industrial plants and new facilities.

Apply the suggestions below and your economic proposals will no doubt be more specific.

The reliability provided by a well-prepared estimate is a vital requirement for the smooth running of any enterprise.

## Guide to making bids

Price competition is part of the buying process, where the buyer invites different tenderers to submit their offer to supply goods or services in a sealed envelope.

The offers must comply with certain technical and commercial requirements specific to each project and previously defined by the buyer.

We define a bidder or proponent as any person or entity that meets the requirements of the purchaser, or the tender documents, and submit the tender or proposal to the competition within the time allowed.

Each of the steps involved in preparing a proposal is analyzed below.

# Bidding steps

The next figure shows the critical steps to be performed by the estimator in the preparation of a bid.

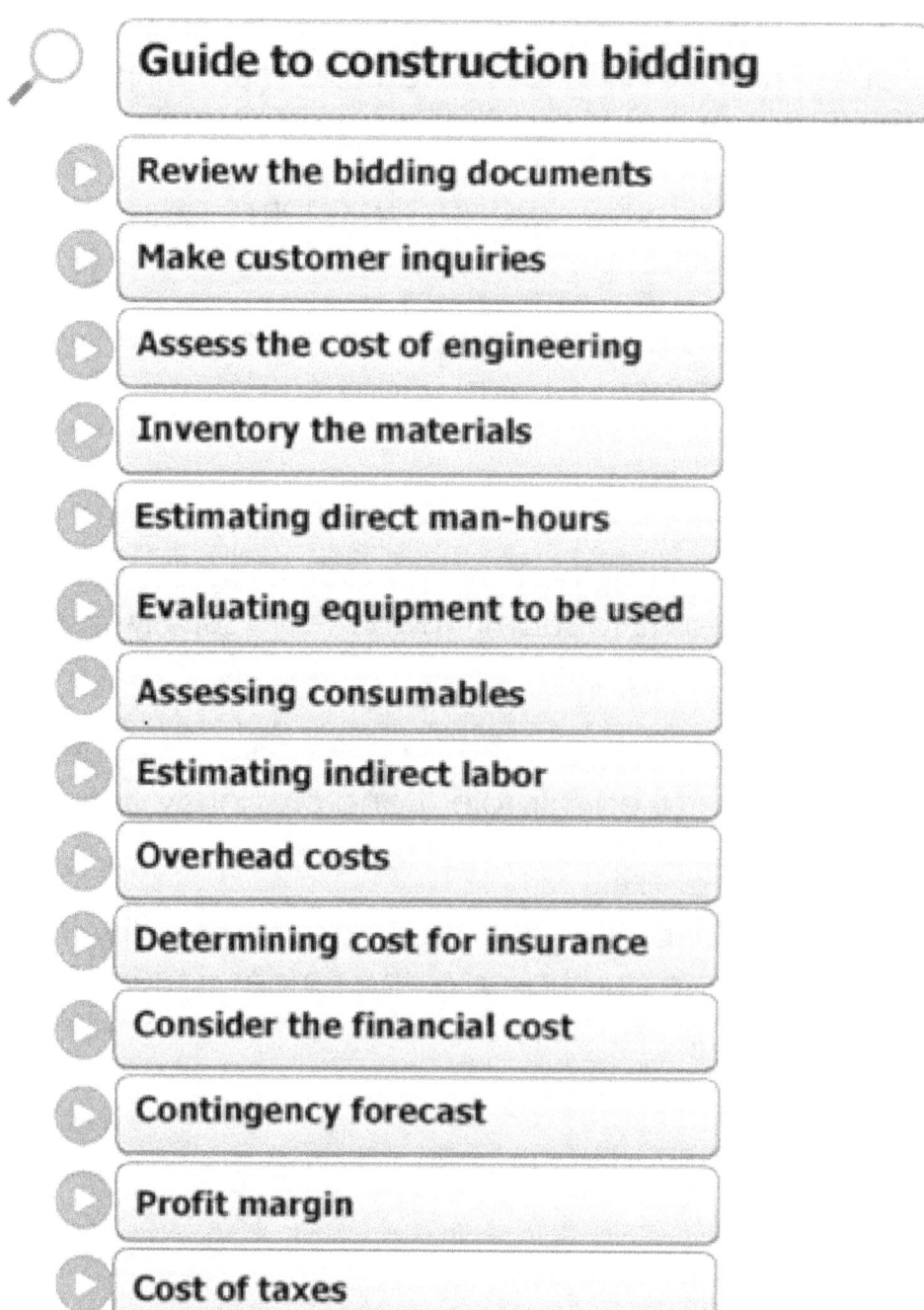

# REVIEW OF TENDER DOCUMENTATION

A bidder's cost and budget estimators must review in detail all attached documents sent by the purchaser in their solicitation.

The review of this documentation is critical and must be reviewed in full before we arrange a visit to the future construction site.

This study raises a series of questions that must be resolved or raised during the site visit.

We should carry the site visits out with experienced personnel who can detect every detail that may affect the performance parameters of the personnel, and the resources that will be used in the course of the work.

The Purchaser must issue a certificate indicating that the Bidder has completed the site visit, which must be attached to the bid.

Failure to submit such a certificate will result in the Bidder's bid being rejected.

Some aspects that should be considered during the site visit are summarized in the following figures.

## Site access

How far will the project staff have to travel from the area where they are staying to the construction site?

Are the access roads passable and properly maintained?

Type of soil at the site

What are the characteristics of the soil on the construction site?

Is it low or muddy and hard to drain, or is it high and dry?

How high is the water table, is the water level stable or variable?

## The climatic condition of the site

To assess the impact of the climate factor on task performance, the following questions need to be answered:

What are the historical weather records of the site?

Does the period in which we execute the work coincide with the rainy, windy, or snowy season?

What are the climatic differences on the construction site compared to the areas where the bidder's personnel usually work?

## Height of the working level

We should review the height above ground level at which each job will be performed to determine, for example:

How much will the execution times increase because of the handling of materials, the movement of equipment and personnel to access the work plane?

Observe if we must add hours for scaffolding assembly.

## Location of the work

When the work is executed within a plant in operation, we must adjust the yields.

The reform or extension of a plant in operation implies entering and to work in the property of a plant in operation, therefore, each activity that is developed in the sector must be agreed with the client. This causes deviations in yields.

The yields to carry out a task are also linked to the characteristics of the plant in operation.

It is not the same to work in an oil refinery that has sectors with a risk of fire or explosion, as to work in a low-risk chemical process plant.

When executing extensions or reforms in operating plants, the estimator must evaluate:

The number and type of joints to be made between new and existing pipelines.

How time-consuming is the task of coordinating activities with the client expected to be?

What delays may be caused by preparing daily work permits, etc.?

## Union's Influence

Evaluate the historical behavior of the local unions.

Find out, for example, how many people of the union's labor pool or in the region are to be employed according to the union agreement.

## Location of workshops

Observe which are the places foreseen for the location of the workshop, materials, and equipment and how far they are from the site carrying out of the work.

Other aspects to observe.

Does the project have a very demanding construction schedule?

If yes, please ask if it is necessary to work at night.

Are the details on the preliminary engineering to quote sufficient, or is it necessary to start a round of consultations?

Determine whether contractors, experienced staff or assistants are available in cities near the workplace.

To accomplish all these tasks, it is essential to carefully read the specifications, or the documentation provided by the purchaser before making the site visit.

How to Ask the Buyer Questions in the Pre-bid

Here, issues that are resolved through consultations with the purchaser are clarified and discussed.

The registration of the site visit and the answers provided by the buyer with the bidders' questions are binding and form part of the tender. The purchaser's responses are included in the offer.

In practice, the buyer opens a communication channel with bidders for a period to respond to questions raised during the quotation period.

Bidders may generally request information from clients in advance in computerized form, but must confirm it in hard copy, duly signed by authorized personnel.

The questions and their answers are sent at the same time in the form of a circular to each tendering company.

The responses shall form part of the tender document.

Two or three days before the bid closing date, all bidders are deemed to have been notified of all changes, provisions, circulars, and replies constituted until this date.

## Proponent's responsibility

It is the Bidder's responsibility to visit the Purchaser's offices to request a copy of all inquiries and responses or access its site to carry out all acts dictated by the purchaser before making the offer.

In other cases, the buyer prefers to call all the bidders to a meeting prior to the close of the bidding process, and distribute an addendum, the contents of which include each of the bidders' questions and the corresponding responses of the purchaser.

In the replies, the buyer shall specify, if applicable, the modifications, modifications, deletions, or additions to be made to each drawing or tender document.

This addendum is to be signed by each bidder as part of the bid document.

# ENGINEERING COST EVALUATION

## Conceptual Engineering

Conceptual engineering is the first phase of an engineering project.

At this stage and once the technical and commercial feasibility of the project is verified according to the objectives set by the client, it is analyzed:

The type of technologies to be used.

The framework of technical standards that will regulate the design.

The economic evaluation and calculation of the profitability of the project.

Then, and with the above data in hand, the client decides whether to continue with the basic engineering of the project or cancel it due to low profitability.

## Basic Engineering

Basic engineering defines the general guidelines and basic ideas of the project.

These ideas and definitions of the project are the pillars on which detailed engineering will be based on the execution of the construction drawings.

On some occasions, the basic engineering of a project is quoted together with its construction.

We should note that the price of this type of engineering is significant regarding the cost of construction, and sometimes may equal or exceed the value of the project's construction cost.

Basic engineering has a high value when:

It is basic engineering with patented technology and specially formulated to execute a transformation process in petrochemical, mining, etc.

In general, these basic engineering packages include:

Process flow diagrams.

Material balances.

Piping and instrumentation diagrams.

Data sheets of equipment and instruments.

Chemical consumption requirements.

Effluent summaries.

## Examples of Basic Engineering

Basic engineering to separate and sanitize industrial effluents from an oil refinery by recycling the hydrocarbons they contain.

The basic engineering converts kerosene into aviation naphtha.

Basic engineering to process iron ore into steel through direct reduction of iron ore, etc.

In other words, the idea, the concept, the technology, the creation exceeds the cost of the action of constructing.

When the professional or the company selling the basic engineering accepts it, the client first acquires the design or the basic engineering that suits his requirements and then tenders the detailed engineering and the construction of the work separately.

In such cases, the vendor shall establish the value of the design or engineering basis.

# Detailed Engineering

Detailed engineering includes manufacturing and/or construction drawings, calculation reports, technical specifications, Data sheets, as-built drawings, etc., and its cost varies according to the type of project.

To estimate the cost of this engineering, it is not only necessary to consider the hours of the specialists, designers, and draftsmen, etc., but it is also necessary to consider the following topics, among others:

Knowing what the client's level of demand is in terms of quality requirements, level of detail, delivery dates, and times for reviewing the documentation that makes up the detailed engineering.

Concerning the above, it is important to note that the cost of engineering increases when the time stipulated by the client to review each drawing and return it with or without observations is greater than usual.

In some projects, especially in cases of extensions or repairs, there are interferences with existing facilities, and we must resolve these with a greater expenditure of hours.

In this type of situation, the cost of detailed engineering, in general, is obtained by comparison with similar works.

In tenders where detailed engineering is not complicated, it suffices to consider the hours of specialists, designers, draftsmen, supplies, etc.

# INVENTORY OF MATERIALS

Steps to identify direct materials, inventory them and obtain information on their costs, delivery times and payment terms.

The term materials comprehend.

-Semi-finished materials which form the basis for the next stage of manufacture, for example, metal profiles from which a structure is to be constructed.

-Cataloged materials.

-Processed materials, such as equipment, which are assembled within the components already manufactured.

In this case, the estimator must identify each of the materials and calculate them, ask for their prices, delivery times and payment terms.

We must complete the above task within a limited time frame that allows the information to be available at the right time to be incorporated into the bid.

The specifications of the required materials are stated in the tender documents or in the documentation provided by the purchaser.

In case of doubt, tenderers may request information and clarification from the purchaser, up to a certain number of working days prior to the submission of proposals.

The contractor's written responses to requests for information and clarifications are, as mentioned above, also notified to all bidders and form part of the tender documents.

# Example of an inventory and estimate of the direct cost of materials at industrial facilities

At this stage, it is necessary to emphasize the importance of carrying out an accurate calculation of the materials required.

Beware, a minimum error in the calculation of expensive material magnifies the error high values.

## Cost of materials

As the materials are identified and counted, the commercial activities of getting an estimate for each material with its delivery times and payment conditions begin.

## Quotation for form

A common and effective practice in developing the quote is to divide the project into categories, sub-categories, and items.

Each item contains the materials and work required for a specific task.

## Examples of categories

Civil works category, piping and equipment category, electrical works category, instrumentation, and control category, before commissioning and commissioning category.

The specialists in each category will divide each of them into subcategories and these into unique items and sub- items, to estimate the direct cost of materials.

# A- Civil works category

*A-1 Subcategory: Reinforced concrete base.*

A-1-1 Item: Excavating and Preparing Resistant Soils.

A-1-2 Item: Cut, fold and arm reinforcing iron.

A-1-3 Item: Constructing and assembling the formwork.

A-1-4 Item: Concrete casting.

A-1-5 Item: Dismounting and cleaning of the formwork.

# B - Category Piping and equipment - Estimate of direct material cost

*B-1 Subcategories: Air Piping.*

B-1-1 Item: Material Identification and Prefabrication Construction.

B-1-2 Item: Hydraulic tests on prefabricated products.

B-1-3 Item: Painting and transfer of prefabricated materials.

B-1-4 Item: Assembling prefabricated sections, supports and change sections.

B-1-5 Item: Last hydraulic test and paint retouching.

# C - Electrical Category

*C-1 Subcategory: Placement of underground conduits.*

C-1-1 Item: Excavations.

C-1-2 Item: Assembly and fixing of conduits.

C-1-3 Item: Lean concrete casting.

C-1-4 Item: Cable assembly and identification according to the technical specifications.

Once the project has been divided into categories, subcategories, and items, we must record the following in each item:

We examine the drawing or documentation of the item.

We compute the material under its specification.

The unit of measurement used.

The data from each supplier.

The unit cost of each material comes from the sum of:

The unit cost of the material at the supplier's premises

The percentage of waste per unit.

The cost of transporting it to the site.

These spreadsheets also record the hours consumed by work teams to complete each activity per unit of measurement.

The next figure shows a typical model for recording and testing tasks.

| Item | Referral drawing | Material identity | Unit of measure | Supplier data | Material cost per unit of measure | Waste by unit of measure | Cost of transporting the material to site per unit of measurement | A Total cost of material put in work per unit of measure |
|------|------------------|-------------------|-----------------|---------------|-----------------------------------|--------------------------|-------------------------------------------------------------------|----------------------------------------------------------|
|      |                  |                   |                 |               |                                   |                          |                                                                   |                                                          |
|      |                  |                   |                 |               |                                   |                          |                                                                   |                                                          |

| Consumption in hours of the typical work crew to execute the unit of measure, obtained from yield tables | Number and specialty of workers that form the crew | Work crew cost | B Direct labor cost per unit of measure | A + B = C Cost of material plus item labor per unit of measure | D Total revised quantity of sub item materials | C * D = E Total direct cost of the sub item | Date and signature of supervisor |
|---|---|---|---|---|---|---|---|
|   |   |   |   |   |   |   |   |
|   |   |   |   |   |   |   |   |

In the figure, we highlight the inventory, and direct cost of materials in a red box.

*Category: Civil Works - Subcategory: A-1 | Reinforced Concrete Base*

Tools to implement this cost.

The most used tool for this task is the Microsoft Excel program, because it allows the use of numerous predefined formulas and templates that save the work.

The Excel worksheet format is created once, then this template is used for other tasks, adapting it to each project.

Excel also enables you to quickly edit calculations or costs.

But if you want to save even more time and avoid human error in data capture, it is convenient to use software to automatically list data, e.g., from AutoCAD.

The advantage of using already integrated applications and tools is that with each change we update all data without having to recover all the information.

# ESTIMATING DIRECT MAN-HOURS

A comprehensive guide to calculating the number of man hours of direct workers on a construction project.

For the estimation of the direct work to be employed in the performance of a work, we demand it at least.

That the person responsible for the estimate or the group of estimators and consultants who make up the working team for this invitation to tender have the experience and capacity to define the logical sequence of execution of each of the stages of construction or assembly of the work being quoted.

Availability of accurate information on the number of man-hours required to execute each project task.

## Labor Calculation Performance Tables

The process of measuring the consumption of man-hours or fraction of man-hours that are necessary to perform a task; requires defining the conditions under which we accomplish the measurement.

This ensures that replication can corroborate the measure.

We describe these conditions as standard conditions.

## Standard conditions

The total man-hours calculated using the records in the tables are only valid if we execute the project under conditions like those existing at the time, we made the performance measurements.

Since each project differs from the other, we must adjust the total hours, resulting from the mathematical calculation with factors that consider the influence of the prevailing conditions in each project.

*In short:*

Estimators must have reliable records of the labor hours to be spent on completing each activity.

Knowing the number and type of tasks in the project, the number of man-hours can be calculated mathematically, which, according to the outputs indicated in the tables are required to complete the undertaking.

Estimators or their evaluators must have full information on the conditions under which we will run each project.

Comprehensive information about the conditions under which we develop each project, as mentioned in the previous paragraph, is obtained from:

Tender documents.

The information obtained during the site visit.

Consultations with the purchaser.

The next step is to define the correction factors to be applied to the total hours of work which have been calculated mathematically.

After estimating the direct hours of work required for each specialization and with the data on hours of work, we obtain the number of workers to be employed.

In short, for this task, the estimator relies on the drawings, the calculation of materials, the tables with performance records, and especially on his experience and the contribution of the specialists.

After estimating the direct working hours required in each specialty and with the date of the work term, we define the quantity and the specialty of the workers.

With respect to the hourly cost of workers, in general, companies have already established the hourly cost of each person dressed and by specialty with workwear, safety features, and toolbox if appropriate.

## Calculation of Hours of Work - Increased Costs of Overtime

Working overtime produces a significant increase in the workers' labor costs, which must be considered in the estimate.

Working overtime is gratifying for the worker, but we should note it that paying for the hour with a 50% surcharge does not increase the worker's performance by 50%; on the contrary the excessive extension of the working day has the consequence of decreasing the average performance of the worker.

The discrepancy between costs per hour versus performance is even more noticeable when the hour is paid at 100%.

Despite the above construction and industrial assembly, companies usually work on their sites for more than 8 hours a day from Monday to Saturday, and

sometimes also with some or all their staff about Sunday considerations and holidays.

It should be noted that the employer plans to do the work with extended working hours for two reasons:

To offer attractive salary compensation to the worker, considering that he has only assured his work continuity of that work and that in addition, the site of work is frequently far from his home.

Balance the decrease in the effective working time of each day due to operational problems such as.

Counteract time lost due to daily mobilization and demobilization of the worker with tools from the workshop to the point of work.
The construction site is often far from the workshop or high above the ground, etc.

Compensation for downtime until work permit approval, etc.
Other situations which lead to an extended working day are:

Works with short lead times and high penalties for noncompliance.

Restoration or expansion of existing facilities on sites requiring rapid development (typically in architectural and civil works).

Work performed by specialized personnel that are not abundant in the labor market, etc.

It should be noted that each country has its labor legislation which, among other things, defines the following issues:

How many hours a week is normally paid according to local labor legislation?

What is the maximum number of authorized overtime hours per day, month, and year?

When are overtime hours paid 50% and 100%?

It should be noted that, regardless of what is regulated by labor legislation, it is usually the approved collective labor agreements of the trade unions that regulate how many hours are worked per week and how overtime is paid.

The conditions laid down by these agreements must meet the condition that they are more favorable to the employee than what is stipulated by national legislation.

The following illustration shows how work hours are calculated.

The table shows the importance of the additional cost for overtime.

It is recommended to consult with the company's accountant to determine the cost of person-hours for a project.

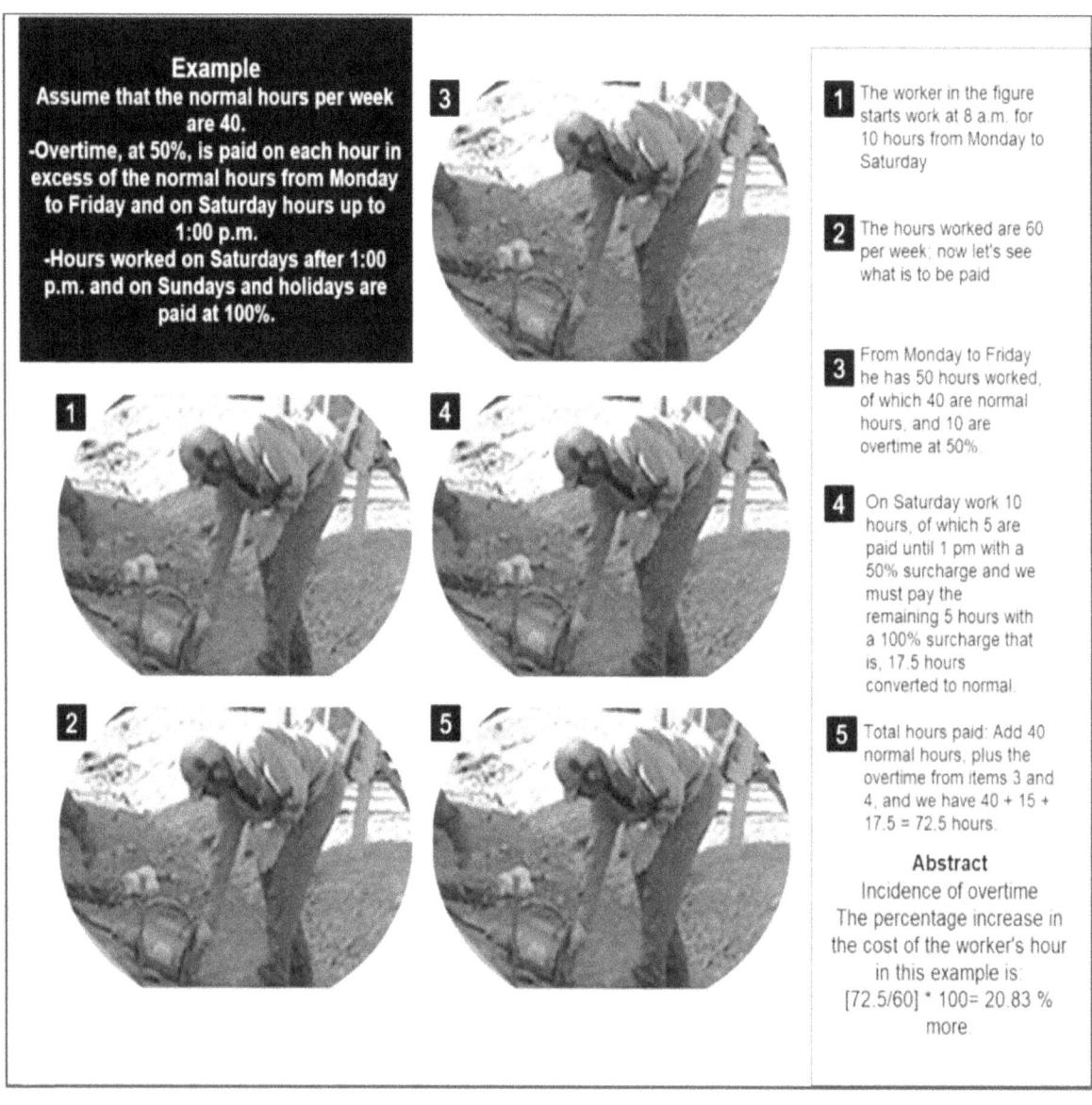

# Example of how overtime affects the calculation of labor hours.

*Hypothetically calculating the impact of overtime*

Assumptions:

We assume that in the area called X, the maximum number of standard hours per week is 40.

50% overtime is paid for every hour that exceeds normal hours from Monday to Friday and Saturday until 1:00 p.m.

Hours worked on Saturdays after 1 pm., and on Sundays and we pay holidays at 100%.

The percentage by which the cost of the hour is increased by the incidence of overtime arises from the ratio of the cost of hours paid over the hours worked per week, multiplied by one hundred.

Therefore, if the workday starts at 8am. and we work 10 hours Monday through Saturday, we have:

# Paid hours each week

Monday through Friday we work 10 hours a day for a total of 50 hours.

This is 10 hours more than the normal 40 hours, so are paid at 50% that are 15 hours converted to normal, (we assumed that the normal maximum number of hours per week is 40).

On Saturday work 10 hours, of which five are paid until 1 pm with a 50% surcharge and we must pay the remaining 5 hours with a 100% surcharge that is, 17.5 hours converted to normal.

Total time paid: (sum of points 1 to 3), equals 40 + 15 + 17.5 = 72.5 hours.

## Hours worked weekly

Monday to Saturday = 10 hours * 6 days, which is 60 hours worked.

## Impact of overtime

The percentage increase in the cost of workers' time is... [72.5/60] * 100 = **20.83% more.**

# ASSESS THE KIND OF EQUIPMENT

To complete this step, the estimation group, with the support of experts, must define the type of equipment required for the project and then establish its costs.

## Type of equipment

The term equipment includes all machines, tools, furniture and their accessories, vehicles, computers, electronic equipment, office machines, etc., used in the performance of the work.

To define the type of equipment required to complete a project, it is analyzed and reviewed, such as:

Specification requirements and associated documentation.

The information analyzed in the previous phases on "Identification and calculation of materials," and "Calculation of direct man-hours."

Descriptions provided by the experts on how the mentioned project is executed step by step.

The preliminary work plan of the project defines the time of use of the machinery, vehicles, etc.

## The preliminary analysis identifies:

The type and quantity of vehicles for the transfer and internal movement of workers and hierarchical personnel (collectives, 4×4 or 4×2 vans, vehicles for supervisors, etc.)

The number and type of trucks for general services, (e.g., those required for the supply of drinking water, irrigation and industrial water, vehicles for the internal transport of materials and machinery, etc.)

The type of heavy and light equipment, mobile cranes, welding machines, motor welders, power-generating units, etc., required to execute the project.

## Calculating the cost per hour for vehicles and machinery

The lifetime of vehicles and machinery = Vu

We define lifetime as the period in which the equipment operates with reliable and economically.

The lifetime of work equipment is affected by the severity of the conditions in which it is used, the care with which it is maintained and repaired, and by its technological obsolescence.

Usually, estimates of useful life in hours of work. To guide the reader, here are the values:

Light construction machines: 6,000 total hours of work; 3 years (e.g., drills, compressors, welding machines, etc.)

Heavy Equipment: 10,000 total work hours; 5-year duration (e.g., front end loader, motor grader, etc.)

Ultra-heavy machinery: 16,000 total work hours; 8-year duration (e.g., asphalt plant)

To reach the above levels, we assumed 2000 hours of work a year, which is close to reality.

## Acquisition value = Vad

The acquisition value represents the market price for the equipment.

To quantify the acquisition value, we should take all costs attached to the acquisition of the equipment into account.

If the equipment is foreign made, the acquisition value of the equipment should include the price of the unit placed in the port of shipment (FOB), the costs of loading, freight, and unloading at the port of destination (CIF), port storage fees, insurance of

goods in transit, other related costs (such as letters of credit, guarantees, etc.), transport to the owner's machine park, etc.

Net value of equipment = $V_n$ = Net value = $V_{ad} - P_4 - P_e$

Where $P_4$ is the value of rims and $P_e$ is the value of special pieces or accessories.

## Resale value of machinery and vehicles at end of life = $V_{rev}$

The recovery value of heavy equipment (loaders, graders, tractors, etc.) typically ranges from 20% to 25% of the acquisition value.

The salvage value of light machinery and equipment (compressors, mixers, motor pumps, motorized welding machines, etc.) generally ranges from 10 to 20% of the acquisition value.

## Value to be depreciated = D

Depreciation is the loss in value of the equipment due to usage or age.

If, as usual, the value of the equipment is assumed to decrease from its original total cost at a uniform rate, the linear method of depreciation is used.

In other words, amortization per hour is the ratio of the acquisition value minus the resale value to the service life of the equipment. This equation is as follows:

*D = [Vn-Vrev]/Vu*

## The average value of the equipment = Vpm

Net asset value plus resale value divided by 2.

*Vpm = [Vn + Vrev]/2*

## Cost of maintenance and repair

Good maintenance extends the economic lifetime of a machine.

Recommended maintenance is remedial, preventative, and predictive.

The cost of maintenance for repairs and spares is normally considered a percentage of the depreciation value in accordance with the following values.

Heavy maintenance cost: 80-100% of amortization value.

Maintenance costs for regular work: 70-90% of depreciation value.

Light vehicle maintenance costs: 50-80% of depreciation value.

We take the cost of labor and the cost of spare parts as a percentage of the maintenance cost of the machinery or vehicle during its useful life.

*The labor cost is 25% of the maintenance cost.*

The cost of spare parts represents 75% of the maintenance cost.

## Cost of fuel use

We link the fuel consumption of building equipment with:

The rated engine power.

The kind of fuel used.

The operating factor of the machine or equipment varies with the speed at which the engine is charged.

The type of operation the machine carries out.

To the expertise of the operator.

Due to technological advancement of equipment, etc.

Normally, each company records the fuel consumption of each unit, for example, for one month, and then divides this amount into the monthly working time of the machine, thus getting the hourly fuel consumption of the equipment.

## Cost of consuming machinery tires due to their use = P4

Heavy machines require special tires for different construction applications. The hourly cost is calculated by dividing the price of the tire by the economy or the useful lifetime of the tire.

Tire service hours are usually provided by the manufacturer and are dependent on the severity of the use.

To the above value is generally added the amount for tire repair, which is taken as 15% of the tire depreciation.

## The hourly cost for assigned personnel

If it's a vehicle, it's their driver. For a concrete installation, it will be its qualified operator and its assistants.

*Interest in capital invested.*

Every company which buys machines finances the funds in the banks or in the capital market, paying the corresponding interest.

This can be done if the entrepreneur has sufficient capital, it makes the investment directly while awaiting that the machine is depreciated in proportion to the investment made.

Here, an amount equivalent to the interest on the capital invested in the machine is charged.

Interest on capital invested annually is calculated as the average value of the equipment (net worth plus residual value divided by 2) by the interest rate.

## Insurance, taxes, and storage expenses

Insurance premiums vary depending on the type of machinery and the risks to be covered during its economic life.

This charge exists both if we ensure the machinery with an insurance company and if the owner is self-insured.

The type of insurance that should be considered is the set of risks. The cost of such insurance is approximately 5.5 per cent of the average value of the equipment.

Taxes are imposed on acquiring property. We must calculate its percentage based on current legislation and may vary from 1% to 2% of the average value of the equipment.

As regards storage, this is the cost of keeping the machinery in the central workshops because of inactivity.

We estimate this cost to be anywhere from 1% to 1.5% of the average value of the equipment.

# Chart of equipment costs

The following chart lists the items that constitute the hourly cost of the equipment

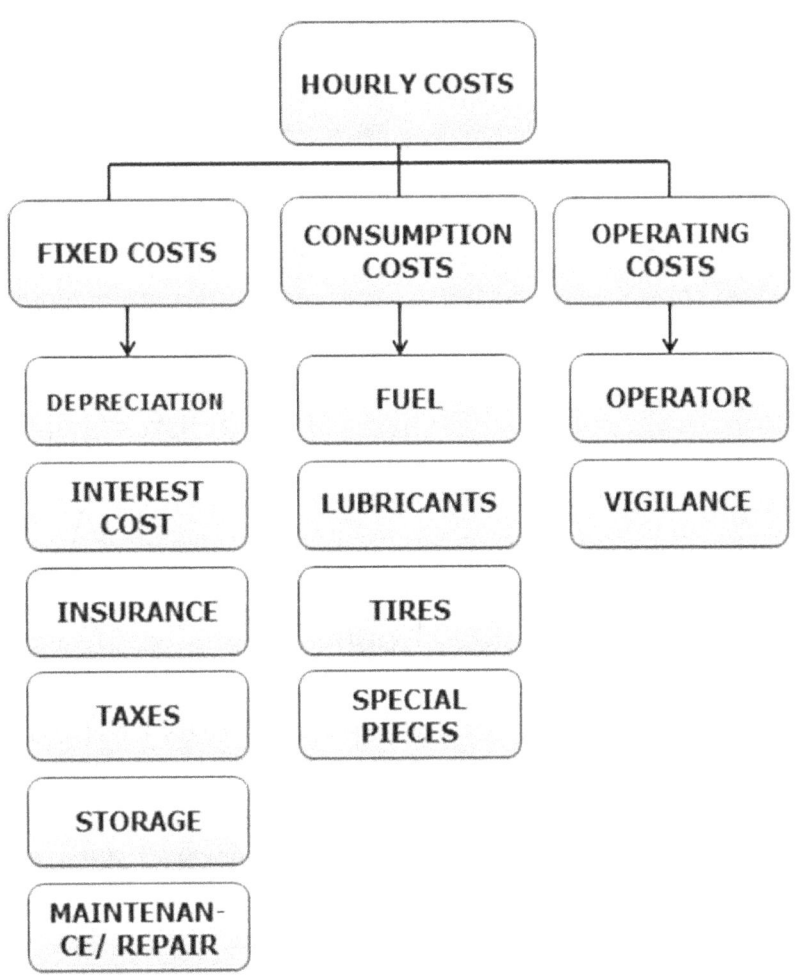

Applying to the above table, we have:

Fixed costs per hour:

Depreciation

Depreciation charge:

$D = [V_n - V_{rev}/V_u]$

Average interest on investment = Im at the annual interest rate = i

The expense to be added to the cost of the equipment for the average interest of the investment is equal to:

The average value of the equipment by the annual interest, divided in 2000 hours per year.

$Im = \{[V_n + V_{rev}]/2 * 2000\} * i$ Cost of insurance, taxes, and storage

Insurance fees, taxes and annual storage fees are as follows:

Insurance cost: 5.5% of the average value of the equipment.

Cost for taxes: 1.5% of the average value of the equipment.

Cost of storage 1.5% to 2% of average equipment value.

Hourly costs are obtained by dividing the amounts by the number of annual work hours, that is, 2000 hours.

Maintenance and repairs of machinery Maintenance, repair, and spares costs.

The cost per hour of maintenance to repair and spares is equal to:

Heavy-duty maintenance costs: 80 to 100% of the depreciation value.

Maintenance expenses for normal work: 70 to 90% of the depreciation value.

Maintenance costs for light jobs: 50-80% of the depreciation value.

# Consumption costs:

*Fuels*

Usually, each company records the fuel consumption, for example, for one month and then divides that data into the monthly working time of the machinery, thus getting the hourly fuel consumption.

The hourly cost of fuel consumption is the product of the expense for the price of fuel.

*Lubricants*

The correct method for finding out the hourly consumption of oil in a machine is to record the engine, hydraulic, and transmission oil consumption for changes and refills at the end of each month and then divide each data into the monthly working time of the machine.

The cost to be taken for oil consumption is obtained by multiplying the hourly consumption of oil by its price.

*Cost per consumption of machinery or equipment tires because of their use.*

The cost per hour is obtained by dividing the price of the tire over the economic life of the tire.

A 15% premium is also charged for their depreciation for tire repair.

# Operating costs:

*Labor costs*

The cost of the worker's hourly wage, plus social security contributions multiplied by the percentage affected, is equal to the hourly cost of the worker.

*Labor cost of surveillance*

The hourly cost of surveillance is equal to 10% of labor costs.

# Total cost

Adding the above fees, we arrive at an estimate of the cost per hour of the equipment.

At the end of the manuscript, a sample application is developed.

Knowing the amount of direct and indirect personnel to be employed in the work and the trips to be made we can estimate the number of vehicles that will be used and their fuel and maintenance costs, etc.

In this phase, quantities, and costs for hand tools to be used (one toolbox per officer is calculated), machinery and special equipment, for example, welding machinery,

generators, optical levels, theodolites, amperometric clamps, etc., should also be evaluated.

The equipment required to transport materials and machinery to the site daily must also be calculated.

# Evaluate Input and Consumption Materials

The term "consumables" refers to all auxiliary materials used for the work, but not incorporated in the construction.

In addition, the contribution material is also an auxiliary material, but it is incorporated into the construction.

This step of the estimation includes all indirect materials used during the execution of work.

# Consumption materials

Consumables are auxiliary materials used in the construction and assembly of a project, but as their name indicates, they are consumed and not detected in the finished project.

Examples of consumable materials:

*Consumption materials in civil works*

Certain civilian work estimators add the following costs to the consumable item:

Waste framing materials, produced by cutting, reuse and handling.

The use of elements for assembling and fastening formworks such as nails, threaded rods, spacers, clips, etc.

Warning: The formwork has a significant value in the cost of m3 of concrete and their cost of purchase and assembly is charged as a direct cost.

*Perishable tools*

For example, drills, cutters, saw blades, steel reamers, cutoff, and roughing discs, etc.

*Consumer materials in electromechanical works*

Oxidizing gases, such as acetylene and oxygen, used in the Oxy - cut.

Shielding gases, such as carbon dioxide, are used in the process of short-circuiting and globular transfer welding.

Traditional binary mixture of argon and carbon dioxide, used in the MIG/MAG process, and gas tube electrode welding.

Special gas mixtures. It can be used as a protective atmosphere in MIG, TIG and plasma welds.

We use carbon dioxide and nitrogen as an inert protective or isolating atmosphere, etc.

# Input materials

Input materials include any indirect material that is incorporated in whole or in part into the finished project.

Examples of input materials.

*In civil works*

The wire to bind the armor.

The spacers between the formwork wall and the reinforcement iron.

Building additives, etc.

*In electromechanical works*

Electric connectors, insulation tape, minor screws, adhesives, etc.

Soldering and brazing filler metals.

Filler materials for electrical arc welding.

*The following table allows you to calculate the number of inputs necessary in each welding process to deposit 100 kg of metal.*

| Process | Consumables per 100 kg of metal deposited | | |
|---|---|---|---|
| | Electrode (kg) | Flux (kg) | Gas (m3) |
| Cellulosic manual electrode | 155 | - | - |
| Manual rutile electrode | 145-170 | - | - |
| Manual electrode low hydrogen | 160-170 | - | - |
| MIG (short circuit) | 110 | - | 17-42 |
| MIG (spray) | 108 | - | 7-11 |
| Tubular with protection | 122 | - | 4-20 |
| Tubular without protection | 126 | - | - |
| Submerged arch | 102 | - | - |

Some filling materials and consumption are paid for as a percentage associated with the volume of what is built, and other higher unit costs like electrodes with their quantity and quality.

*Table showing the weight of metal deposited for different joints per linear meter.*

| Table with the weight of the weld metal deposited for different welded joints. Valid for steel | | | | | | |
|---|---|---|---|---|---|---|
| Type of joint | ⌐ | ╞═╡ | ∨ | ✕ | ╱╲ | ◁ |
| E (mm) | Metal deposited (kg/ml) | | | | | |
| 3,2 | 0,045 | 0,098 | | | | |
| 6,4 | 0,177 | 0,190 | 0,380 | | 0,358 | |
| 9,5 | 0,396 | | 0,638 | | 0,605 | |
| 12,5 | 0,708 | | 1,168 | | 1,066 | |
| 16 | 1,103 | | 1,731 | | 1,707 | 1,089 |
| 19 | 1,592 | | 2,380 | 1,049 | 2,130 | 1,449 |
| 25 | 2,839 | | 3,987 | 2,578 | 3,554 | 2,322 |
| 32 | | | | 3,768 | | 3,380 |
| 37,5 | | | | 5,193 | | 4,648 |
| 51 | | | | 8,680 | | 7,736 |
| 63,5 | | | | 13,674 | | 11,617 |
| 76 | | | | 18,432 | | 16,253 |

Warning: When working with many welding inches or with alloy or special steels, the cost of these materials is significant.

# ESTIMATION OF INDIRECT LABOR

## How to Calculate Indirect Labor in Construction.

Labor that is not assigned to a specific task or process is called indirect labor.

The total indirect labor budget in a construction company is the sum of the indirect hours for each work in progress plus the indirect hours in the central office for engineering, budgeting, quality control, purchasing, sales, administration, accounting, finance, services, etc.

In this phase, only the indirect work of the staff assigned exclusively to the project mentioned is considered.

## Generalities for estimating indirect labor

One aspect to consider is that the costs of personnel driving machinery and vehicle drivers have already been charged at the "equipment cost" stage.

In turn, the costs of personnel inputs categorized as indirect were also charged in the "Valuation of consumables" stage.

How to calculate the indirect work of a project

Indirect labor costs group wages paid to workers who perform tasks that are not directly involved in actively converting materials into finished products or providing services.

For instance, some considerations in estimating indirect labor costs in a project are described:

*Supervising or supporting personnel*

The site supervision group includes the project lead, foremen, supervisors, etc.

The work management supervises the work.

director who is the professional person in charge of the supervision, monitoring, and control of the project.

The tasks carried out by the project manager are multiple and, depending on the complexity of the project, may or may not require the assistance of an entire multidisciplinary team.

*Principal tasks of the Project Manager:*

Verify that the resources provided to execute the project are sufficient to execute the work plan in time and form.

Control, through foremen and supervisors, that the work is performed in accordance with the requirements of engineering plans, specifications and expected performance.

Verify that their personnel and those subcontractors are following the quality and safety standards set out in the tender documents.

Complete timely and formal certification of work completed to obtain payment.

Verification of compliance with current labor requirements.

Ensure compliance with environmental standards, etc.

Internal Services Department

This department ensures the timely delivery of core services for every work, such as:

Supplying industrial water of the right quality for sanitary use, cleaning, concrete, testing, etc. We get that supply wells.

Provide potable water to staff, usually through hot and cold-water dispensers.

Supply of electrical energy (if requested by the client in the tender specifications), for example with clean or rented generators.

Maintenance of bathrooms for employees, the best option is always the chemical toilet with wash basins.

Maintenance of staff transport vehicles that transport workers from their homes to and from the construction site.

To achieve this, we use minibuses with less than two years of use and speed controllers. We emphasize that the authorization for its operation is issued by the security service.

Maintain telephone communication system and internet.

Labor to maintain construction machinery and provide fuel and lubricants.

Another daily task is to transfer and remove any work equipment from the site.

*The figure illustrates the different factors that should be considered when estimating indirect costs*

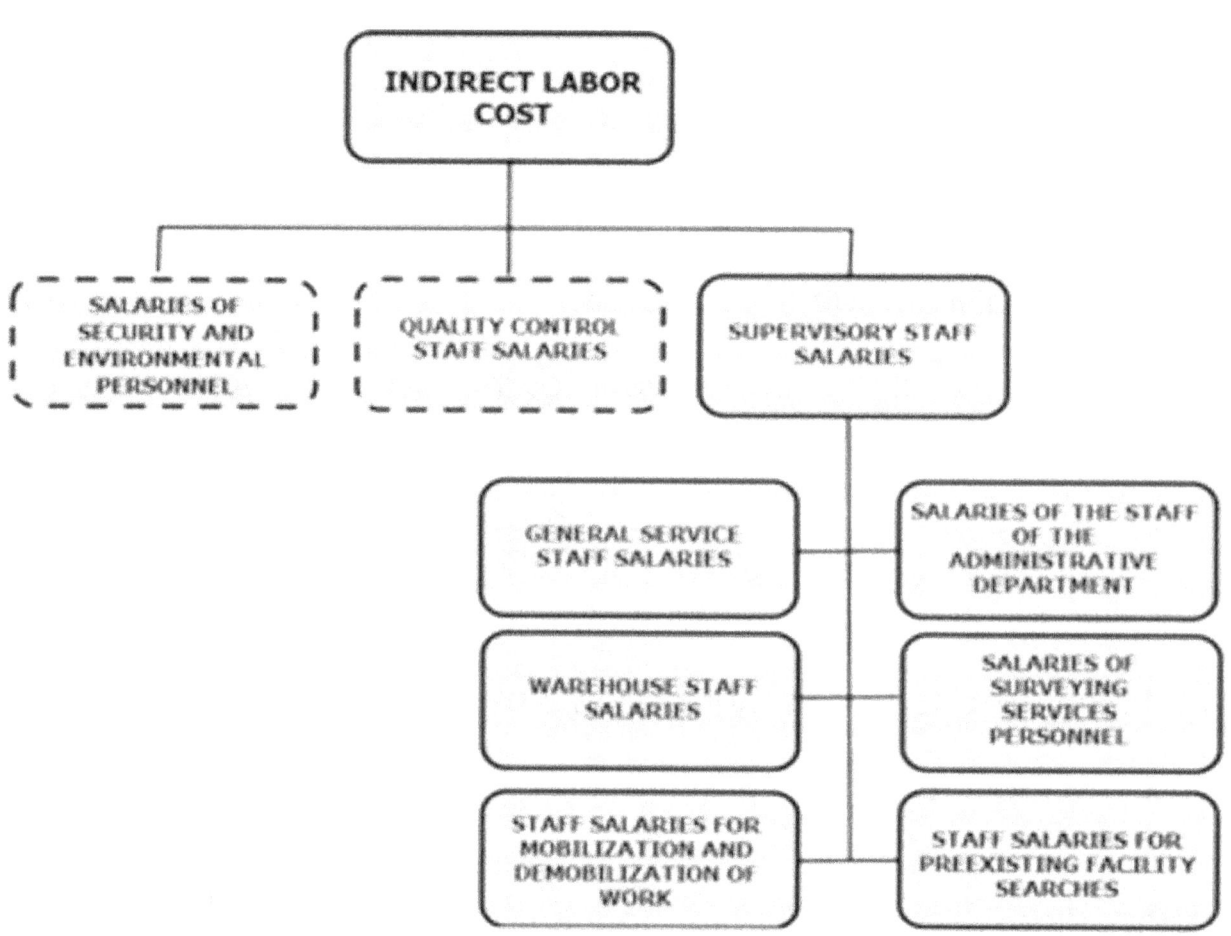

*How to Calculate Indirect Labor costs of the surveying service*

The topography specialist and his employees must ensure the correct positioning and leveling of all components to be built or assembled.

Indirect Labor Cost Examples in Construction - Storerooms

They oversee the management of construction materials, input and consumption materials, spare parts, stock control, ordering and protection of tools, issuing reports, etc.

## Administrative Department

The administrative department controls the documentation of personnel income, the number of hours worked, the completion of payments on the salary accounts, minor purchases, etc.

Independent groups not dependent on construction management.

## Safety and Environment Department

The Department of Security and Environment is an autonomous group responsible for ensuring the security and conservation of the environment, and verifies every task performed on site daily to prevent incidents and/or accidents.

The safety department is the one that decides how a job is performed safely, so that department must approve the procedure with which any task is performed.

## Quality Control Department

This is another independent department that monitors:

Performing tests and samples in civil engineering works.

Verification of pipe traceability for material certification and non-destructive testing.

Verify compliance with the technical requirements of the Technical Data Sheets for equipment, instruments, electrical panels, and each material entering the site.

This department is autonomous and is subject to quality control of the company's central office.

## How to Estimate other Indirect Costs Associated with the Project

Analysis of the items to which the estimator assigns a defined cost.

The magnitude of these costs can be significant and will depend on the characteristics of the project, its location and the requirements identified in the tender documents.

We categorize some of those costs into logistic costs. For instance:

Assembly and disassembly of workshops.

Mobilization and demobilization construction sites.

Search for existing underground facilities.

Costs for setting up and dismantling workshops, and those corresponding to mobilization and demobilization sites are components of the project schedule to which the estimator assigns its identified costs.

Also loaded in this phase is the indirect work to be consumed on the construction site during the work in

the services of cleaning, serene, day supervision, maintenance, etc.

# Layout of a typical workshop

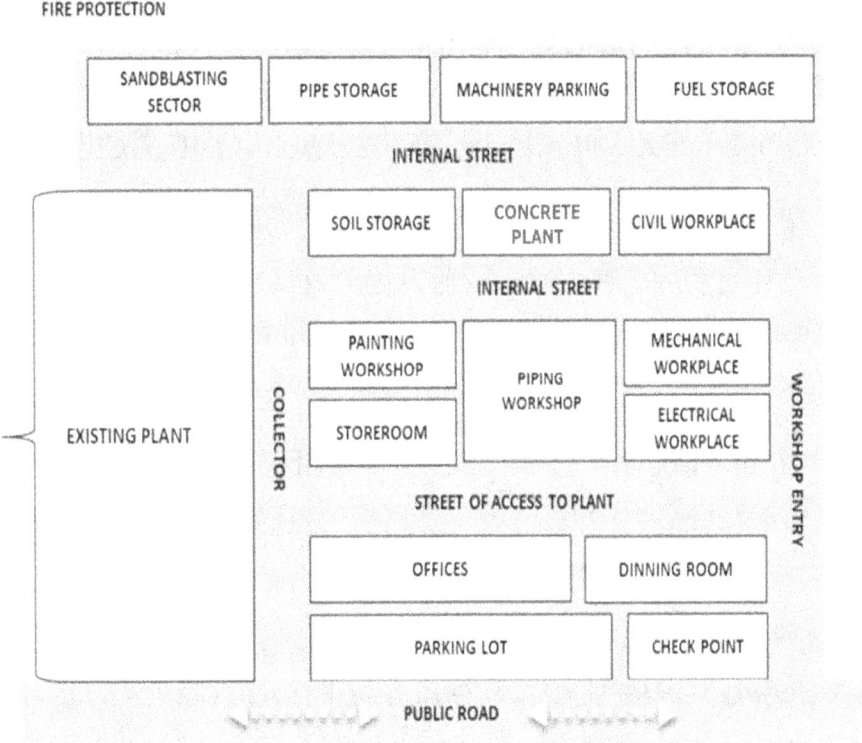

In the layout, we can see, for example, how to organize the various sectors of the workshop to achieve wise use of the area.

We usually agree on the arrangement of the workshop through meetings between interested parties, i.e., the customer, the contracting company and the safety and environment managers.

In the event of an expansion of the plant, the operators participate in the meetings.

In areas designated offices, containers are properly equipped for contractor and client inspection.

In general, sandblasting and painting areas are protected, sideways to prevent contamination, and the prefabricated area is a shield that protects workers from adverse weather.

Perimeter areas free of combustible materials are left protected from fire.

Security and environmental personnel oversee the maintenance of order and cleanliness in the construction site area.

## Investigation of existing underground installations

Another specific element in the project schedule to which the estimator assigns a defined value is the expense incurred when it is necessary to locate pre-existing buried facilities.

In general, it is common for preexisting structures to have signaling defects on buried elements, this often happens to cathodic protection cables, grounding cables and irrigation pipes, and to a lesser extent for process pipes.

When a plant expansion or renovation is to be executed and before the start of fieldwork, we should review all information on the plant in operation to verify the location of existing buried facilities to define what interference may occur.

To release sectors where there is a suspicion of interference because of the existence of poorly marked buried materials, a manual trench is made, arranged in grids, to the depth where the new elements are to be placed.

This trenching in places with the risk of fire or explosion is executed with anti-spark shovels.

Metal detectors, digital electrical current detectors, etc., are also used to determine the position and depth of metal barriers and underground power and signal cables.

We convert all records from these explorations into drawings that are signed by the parties to the compliance signal before any underground activity begins in the area.

# OVERHEAD COSTS

Overheads are indirect costs that remain constant regardless of the company's production level.

This book examines the overhead costs of construction and industrial assembly companies, which provides services for the execution of:

New industry projects.

Modification and/or expansion of existing industrial facilities.

Engineering services.

Work-site management.

Inspection services, etc.

## How to Calculate Overhead Costs in Construction Projects

Overheads costing in Construction Projects, refers to current business costs that are not directly related to the execution of an industrial building or assembly project.

Identifying overhead costs is not only important for budgetary purposes, but also to determine how much a company should charge for the overhead of its new projects to achieve a profit.

## Where do overhead costs come from?

Overheads costing. Construction companies calculate the general expenses of their projects through the evaluation of the 4 items mentioned below.

*First, consider all technical, admin, and maintenance staff salaries (not construction projects)*

Examples: this includes the company's indirect hours of engineering, budgeting, quality control, purchasing, sales, administration, accounting, finance, legal, cleaning, security, etc., departments.

The image shows a summary prepared by the author regarding typical general expenses for construction and / or assembly projects.

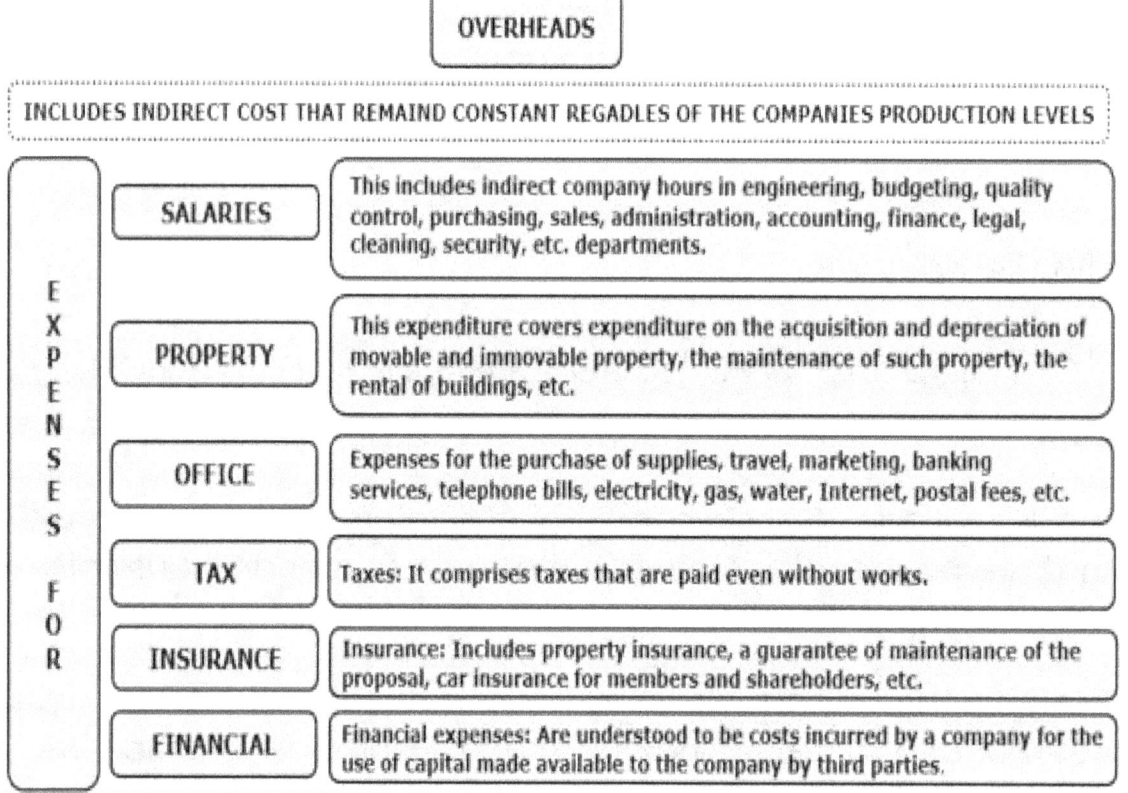

*Then, add the expenditures related to movable and immovable property (not construction projects)*

Examples: this expenditure covers the acquisition and amortization of movable and immovable property, the maintenance of such property, the leasing of buildings, etc.

*Also, add office expenditures (no construction project)*

Examples: expenses for the purchase of supplies, travel, marketing, banking, phone bills, electricity, gas, water, internet, postage, etc.

*Finally, include taxes, insurance, and financial costs (no construction projects)*

**Taxes**: This includes taxes payable even without work.

**Insurance, Examples**: Includes property insurance, proposal maintenance coverage, car insurance for members and shareholders, etc.

**Financial costs, Examples:** Financial expenses are the costs a company incurs for the use of capital provided to it by third parties.

## FAQs:

What is the best method of estimate overheads in a small construction company?

Firstly, based on the 4 elements listed above, we can make a complete list of our overhead costs over a period.

The result gives us the general expenses of the construction company for that project.

*How can overhead be allocated on a new project?*

Based on the type of work performed by the Contractor, they may be distributed as follows:

Allocate overhead costs based on the number of direct labor hours for each project.

Allocate indirect costs based on the material costs of each project.

Allocate indirect costs, according to the square meters of the project.

Assign overhead costs as a percent of direct costs incurred for each project.

What is the percent of a construction company overhead over a certain period?

To obtain the percent of overhead, divide the above total of total sales for this period and multiply by 100.

*The overhead cost formula is:*

> [Total overhead over a period / Total sales over the same period] *100 = Percentage of overhead.

This percent represents what the estimator uses to assign overhead to the new budgets.

On average, construction overhead can range from 5 to 15 percent of direct costs.

However, it is important to note that the actual overhead percent can vary significantly between different construction companies and projects. It is essential that construction companies carefully analyze and manage their overhead to ensure their profitability and competitiveness.

*How to reduce overhead costs? / How to Calculate Overhead Costs on Construction Projects*

Reducing overhead costs is a crucial aspect for construction contractors to be competitive and increase their profit margins.

Here are some specific ways to achieve this in the construction industry.

**Firstly: Use common sense**. Reduce overhead. If we use common sense, we can achieve an optimal level of overall spending.

Reducing general expenses to what is strictly necessary is a crucial step to maintain the financial health of the company.

**Rent your office space.** At the beginning of any business, it is advisable to work with minimal overhead, until income justifies a larger investment.

**Regular Financial Analysis**. Conduct regular financial analysis to identify areas of high overhead costs and implement strategies for improvement.

**Realistic Bidding**. Be realistic in your bidding process to avoid underestimating costs and facing a financial strain during the project.

**Technology Adoption**. Reduce overhead. Utilize construction management software and digital tools to streamline processes, reduce paperwork, and enhance communication.

**Labor Productivity**. Monitor labor productivity and identify areas where improvements can be made to complete tasks more efficiently.

**Employee Training**. Ensure your personnel are well-trained and skilled to reduce errors, rework, and accidents. Well-trained employees work more efficiently.

**Office Expenses**. Look for ways to reduce office-related costs, such as paper usage, printing expenses, and office supplies.

**Benchmarking**. Compare your overhead costs to industry benchmarks to identify areas where your costs may be higher than average and take actions to improve those aspects.

By implementing these strategies and regularly reviewing their overall costs, construction contractors can streamline their operations, improve profitability, and remain competitive in the industry.

Note

Some companies take risks by reducing the amount allocated to overhead in a budget to present a more competitive offer.

This happens when the depreciation cost of heavy machines used in construction is lowered.

# CALCULATION OF THE COST OF INSURANCE

The estimated financial costs in an economic proposal include the cost of insurance and the actual financial cost.

## Insurance Requirements for Bid / Types of Construction Insurance

*1. Determine what types of coverage the client requires*

Tender documents should set out the type of insurance required by the customer, coverage, and insured amounts. The type of insurance to be included in the proposal is divided into tender guarantees and guarantees necessary for the performance of the contract.

We must ensure that there is no lack of any other insurance required under current labor laws. The type of insurance currently used in the construction sector is summarized in the figure below:

**TYPES OF INSURANCE TO BE CONSIDERED IN THE PROPOSAL**

**Coverage to be contracted for bidding**
- Bid bond insurance

**Insurances required at the beginning of the work or project**
- Workplace accident insurance
- Worker's life insurance
- Civil liability for constructions
- Professional liability insurance
- Contractors pollution liability (CPL)
- Business Vehicle and Commercial Auto Insurance
- Contractors Equipment Insurance
- Guarantee insurance to ensure compliance with the contract
- Surety insurance to replace the retention with repair funds
- Policy of guarantee for financial advance and/or collection
- Construction and erection insurance during the execution of a project. All Risks
- Any other insurance required by current labor legislation

## 2. Confirm which insurance companies are accepted by the client

In most cases, in the tender documents, the buyer lists the authorized insurers. If the information is not included in the tender documents, we should submit a written request to the client.

## 3. Request quotation from authorized insurance companies

After defining all the above, the final step is to ask for quotes from various insurance companies to choose the most convenient.

For the Insurance Company's choice, not only should the price be evaluated, but also the certainty that the coverage includes all the risks to be covered. To do so, it is convenient to seek the advice of a risk manager and an insurance consultant.

# FAQs:

### What is Policy/Coverage/Premium?

**Policy:** The policy is the document which gives the validity of the insurance agreement between the insured and the insurer. It sets out the requirements, rights and obligations of the parties involved.

**Coverage:** The insurance coverage is the commitment assumed by the insurer to pay an indemnity to the insured (or his/her beneficiaries), to repair the consequences of a loss.

It should be noted that the coverage has a limit known as the insured capital.

**Premium:** The insurance premium is the price of insurances, that is, the amount of money that the insured pays periodically to the insurance company for the coverage he receives by the insured risk.

Payment of the premium within the prescribed period obliges the insurer to comply with the service agreed with the insured.

## What is a bid bond?

Bid Bond Insurance, also known as a Bid Bond Guarantee, is a type of insurance which guarantees the contractor's commitment to a construction project.

It is generally necessary in the tendering process for both public and private construction projects.

When a contractor submits a bid for a project, they may be required to provide a bid bond as a form of security to the project owner or developer.

The bid bond protects the owner in case the contractor is awarded the project but fails to enter a contract or fails to provide the required performance and payment bonds.

If the Contractor fails to meet its obligations following project award, the Project Owner may claim the Bid Bond.

The insurer will then compensate the owner until the amount of the surety.

The contractor is usually responsible for reimbursing the insurer for any claims paid. Bid bond insurance provides financial protection to project owners, ensuring they have recourse if a successful bidder does not act on the contract.

It helps maintain the integrity of the bidding process and provides assurance to project owners that contractors take their bids seriously.

The bidder assigns the cost of these policies to overhead.

*Insurance required for contractors: Workplace accident insurance.*

Definitions of Workplace accident insurance  Workplace accident insurance, also known as workers' compensation insurance, is a type of insurance coverage that provides financial protection to employees who suffer injuries or illnesses because of their work.

It is designed to ensure that employees receive necessary medical treatment and wage replacement benefits, while also protecting employers from lawsuits related to workplace injuries.

*Here are some key points about workplace accident insurance:*

**Coverage:** Workplace accident insurance covers employees for injuries or illnesses that occur during their employment.

It typically includes accidents, such as slips and falls, as well as occupational diseases or illnesses that develop over time due to work conditions.

**Medical Expenses:** The insurance policy typically covers medical expenses related to the treatment of work-related injuries or illnesses.

This can include hospitalization, doctor's visits, medications, surgeries, rehabilitation, and other necessary medical treatments.

**Wage Replacement:** If an employee is unable to work due to a work-related injury or illness, workplace accident insurance provides wage replacement benefits.

These benefits usually cover a portion of the employee's lost wages, typically a percentage of their pre-injury earnings, and are meant to help the employee meet their financial obligations during the recovery period.

**Disability Benefits**: In cases where the work-related injury or illness results in a permanent disability or impairment, workplace accident insurance may provide additional disability benefits to compensate for the loss of earning capacity.

**Legal Protection:** By providing workers' compensation coverage, employers are generally protected from being sued by employees for workplace injuries or illnesses.

Workers' compensation is considered a no-fault system, meaning that employees are entitled to benefits regardless of who was at fault for the accident.

**Mandatory Requirement:** Workplace accident insurance is often a legal requirement for employers in many jurisdictions. Employers are typically required to carry workers' compensation insurance to provide coverage for their employees in case of work-related injuries or illnesses.

*Insurance required for contractors: Life insurance for workers.*

Life insurance for workers, also known as group life insurance or employer-sponsored life insurance, is a type of life insurance coverage provided by employers to their employees.

It is a benefit offered as part of an employee benefits package, and its purpose is to provide financial protection to employees and their families in the event of the employee's death.

Here are some key points about life insurance for workers:

**Coverage:** Life insurance for workers provides a death benefit to the designated beneficiaries if the insured employee passes away during the coverage period.

The death benefit is typically a lump sum payment and is intended to provide financial support to the employee's family, such as covering funeral expenses, paying off debts, replacing lost income, or funding education expenses.

**Group Coverage:** This type of life insurance is provided to a group of employees rather than individual policies for each employee.

The employer negotiates a group life insurance policy with an insurance provider, and all eligible employees are automatically covered under the policy.

**Employer-Paid or Employee-Paid**: The cost of the life insurance coverage can be either fully paid by the employer as an employee benefit or shared between the employer and the employee through payroll deductions.

**Simplified Underwriting:** Group life insurance typically involves simplified underwriting compared to individual life insurance policies.

This means that employees may not be required to undergo a medical examination or provide detailed health information to qualify for coverage.

However, there may still be certain eligibility criteria, such as minimum hours worked or a waiting period before an employee becomes eligible for coverage.

**Portability and Conversion:** Depending on the terms of the policy, employees may have the option to continue their life insurance coverage if they leave the company.

This is often referred to as portability. Additionally, some policies may allow employees to convert their group coverage into individual policies upon leaving employment.

**Benefit Amount:** The amount of coverage provided under group life insurance is usually a multiple of the employee's salary, such as one or two times the annual salary.

However, employers may offer different coverage options or allow employees to purchase additional coverage through voluntary contributions.

**Tax Considerations:** In many countries, the premiums paid by employers for group life insurance are often tax-deductible expenses.

Additionally, the death benefit received by beneficiaries is generally tax-free.

*Insurance required for contractors: Civil responsibility for construction.*

Civil responsibility in the context of construction refers to the legal obligation of construction professionals or entities to compensate for any damage, injury, or loss caused to third parties because of their construction activities.

Construction projects involve various parties, including contractors, architects, engineers, and subcontractors, and each of them may have civil responsibilities depending on their roles and contractual agreements.

**Duty of Care**: Construction professionals have a duty of care to ensure that their actions do not cause harm to others.

They are expected to follow accepted industry standards, exercise reasonable skill and care, and comply with relevant building codes, regulations, and safety guidelines.

**Negligence:** If a construction professional fails to meet the expected standard of care, resulting in damage or injury, they may be held liable for negligence.

Negligence occurs when there is a breach of the duty of care, causing foreseeable harm to others.

**Types of Damages:** Civil responsibility in construction can involve a wide range of damages, including property damage, personal injury, economic loss, and loss of use.

For example, if a construction defect leads to a building collapse causing injuries to occupants, the responsible party may be required to compensate for medical expenses, property damage, lost income, and pain and suffering.

**Contractual Obligations:** Civil responsibility in construction is often governed by contractual agreements.

Contracts may specify the standards of performance, responsibilities, and liabilities of each party involved in the construction project.

Parties can be held accountable for breaching their contractual obligations and may be required to compensate for resulting damages.

**Professional Liability Insurance**: Construction professionals, such as architects and engineers, typically carry professional liability insurance, also known as errors and omissions (E&O) insurance.

This insurance provides coverage for damages resulting from professional negligence or errors in design, specifications, or construction supervision.

**Statutory Requirements**: Construction projects are subject to regulatory requirements and building codes set by local, regional, or national authorities.

Failure to comply with these requirements can result in civil liability, as well as legal penalties or fines.

**Dispute Resolution:** In the event of a construction-related dispute, the parties may resort to alternative dispute resolution mechanisms, such as mediation or arbitration, to resolve their differences without going to court.

These mechanisms can help parties reach a settlement and avoid protracted litigation.

*Insurance required for contractors: Professional liability insurance.*

It may be applied to various sectors.

Professional liability insurance, also known as errors and omissions (E&O) insurance, protects professionals from financial losses resulting from claims of negligence or inadequate work. It is designed to provide coverage for professionals who provide advice, services, or expertise to clients.

Here are some key points about professional liability insurance:

**Coverage:** Professional liability insurance covers professionals against claims made by clients or third parties alleging errors, omissions, negligence, professional misconduct, or failure to deliver promised services.

It typically includes legal defense costs, settlements, and judgments associated with such claims.

**Professions Covered:** Various professions can benefit from professional liability insurance, including but not limited to accountants, consultants, architects, engineers, real estate agents, etc.

**Types of Claims**: Claims covered by professional liability insurance can include malpractice, negligence, misrepresentation, violation of good faith and fair dealing, and failure to meet professional standards.

**Importance:** Professional liability insurance is important because even professionals who excel at their work can make mistakes or face baseless claims.

A single claim can result in significant financial losses, reputational damage, and legal expenses, which can be financially devastating without adequate insurance protection.

**Tailored Coverage**: Professional liability insurance policies can be tailored to the specific needs of different professions and industries.

The coverage limits, deductibles, and policy terms can vary depending on factors such as the nature of the work, the size of the business, and the level of risk involved.

**Legal Requirements:** In some professions, professional liability insurance is a legal or regulatory requirement.

**Claims Process**: In the event of a claim, the insured professional must notify the insurance company promptly.

The insurance company will typically investigate the claim, provide legal defense if necessary, and negotiate settlements or represent the insured in court, depending on the circumstances.

**Exclusions:** Like any insurance policy, professional liability insurance has certain exclusions.

Common exclusions may include intentional wrongdoing, criminal acts, bodily injury, or property damage claims (covered by general liability insurance), and claims related to other types of insurance such as workers' compensation.

*Insurance required for contractors: Contractor's pollution liability (CPL)*

Definitions of Contractors pollution liability (CPL) insurance.

Contractors' pollution liability (CPL) insurance is a type of insurance coverage specifically designed to protect contractors and subcontractors against liabilities arising from pollution-related risks and incidents.

It provides coverage for damages, cleanup costs, and legal expenses associated with pollution incidents that occur during construction or contracting activities.

Here are some key points about Contractors' Pollution Liability (CPL) insurance:

**Coverage**: CPL insurance offers protection for contractors in the event of pollution-related incidents, such as accidental releases of pollutants, environmental damage, or contamination caused by their work activities.

It covers both sudden and gradual pollution events.

**Scope of Coverage:** CPL insurance typically covers various types of pollution, including pollution caused by hazardous materials, mold, asbestos, lead, and other contaminants.

It may also provide coverage for third-party bodily injury, property damage, and cleanup costs resulting from pollution incidents.

**Covered Parties:** CPL insurance is typically purchased by contractors and subcontractors involved in construction, renovation, or remediation projects.

It may be applicable to various sectors, including general contractors, environmental remediation contractors, construction managers, and specialty trade contractors.

**Policy Features:** CPL insurance policies can be customized to meet the specific needs of contractors.

Coverage limits, deductibles, and policy terms can vary depending on factors such as the size of the contractor's operations, the types of projects undertaken, and the level of risk involved.

**Exclusions:** CPL insurance policies may have certain exclusions, such as intentional pollution, known pollution conditions, and certain types of professional services.

It's crucial for contractors to review their policy carefully and understand the specific exclusions and limitations.

**Contractual Requirements:** In some cases, contractors may be required to carry CPL insurance as a contractual obligation.

Clients or project owners may include it as a prerequisite for awarding a contract, particularly for projects involving environmental risks or sensitive locations.

**Claims Process:** In the event of a pollution incident, the contractor must notify the insurance company promptly.

The insurance company will investigate the claim, provide legal defense if necessary, and cover costs associated with the incident, including cleanup, remediation, and potential liabilities.

**Risk Management**: Contractors should also implement effective risk management practices to minimize the likelihood of pollution incidents and associated liabilities.

This may include proper handling, storage, and disposal of hazardous materials, adherence to environmental regulations, and proactive safety measures.

*Insurance required for contractors: Business and commercial motor vehicle insurance.*

Business and commercial motor vehicle insurance provides coverage for vehicles used for business purposes.

Whether you have a single vehicle or a fleet of vehicles, this type of insurance is designed to protect your business against financial losses resulting from accidents, theft, or other damages involving your vehicles.

Here are some key points to understand about business and commercial motor vehicle insurance:

**Coverage**: Commercial motor vehicle insurance typically provides coverage for liability, physical damage, and other specific risks associated with business-related vehicle use.

**Liability coverage:** This protects your business if you or your employees cause injury or property damage to others while operating the vehicles for business purposes.

It covers legal expenses, medical costs, and property repairs or replacement.

**Physical damage coverage:** This includes comprehensive and collision coverage. Comprehensive coverage protects against non-collision incidents such as theft, vandalism, fire, or natural disasters.

Collision coverage provides protection for damages resulting from collisions with other vehicles or objects.

**Uninsured/underinsured motorist coverage:** This covers you and your employees if you're involved in an accident with a driver who is at fault but has insufficient or no insurance.

**Medical payments coverage:** This covers medical expenses for you and your passengers, regardless of who is at fault for the accident.

**Cargo coverage:** If your business involves transporting goods, you may need cargo coverage to protect the value of the goods in case of damage or theft.

**Non-owned vehicle coverage:** If your employees occasionally use their personal vehicles for business purposes, this coverage can protect your business if they're involved in an accident while doing so.

**Vehicle Types:** Business and commercial motor vehicle insurance can cover a wide range of vehicles, including cars, trucks, vans, delivery vehicles, buses, taxis, and specialized vehicles like construction and assembly machinery.

The insurance can be tailored to suit your specific needs based on the types of vehicles your business uses.

**Premiums:** The cost of commercial motor vehicle insurance premiums depends on various factors such as the number and type of vehicles, their usage, the driving records of the employees, the location of your business, and the coverage limits you choose.

Insurance companies may also consider the industry in which your business operates, as certain industries may have higher risk profiles.

**Legal Requirements:** Business and commercial motor vehicle insurance is often legally required, especially for commercial vehicles.

The specific requirements can vary by jurisdiction, so it's essential to understand and comply with the regulations in your area.

**Risk Management**: Maintaining a good safety record, implementing driver training programs, and properly maintaining your vehicles can help reduce the risks associated with commercial vehicle operations.

Insurance companies may offer discounts or incentives for businesses that demonstrate effective risk management practices.

Insurance required for contractors: Contractors Equipment Insurance.

Definitions of Contractors Equipment Insurance  Contractors Equipment Insurance is a type of coverage specifically designed to protect contractors and construction companies against financial losses arising from damage, theft, or loss of their equipment and tools.

Contractors often rely on various types of equipment and machinery to carry out their work, and these assets can be costly to repair or replace if they are damaged or stolen. *Contractors Equipment Insurance typically provides coverage for:*

**Owned equipment:** This includes coverage for the contractor's own equipment, such as excavators, bulldozers, cranes, generators, power tools, and other machinery.

**Rented or leased equipment**: If contractors rent or lease equipment for their projects, this coverage can protect them from losses associated with damage or theft of the rented or leased items.

**Equipment in transit**: Coverage can extend to equipment and tools while they are being transported to and from job sites.

**Equipment breakdown**: This coverage can help cover the costs of repairing or replacing equipment in the event of mechanical or electrical breakdowns.

**Loss of income**: In case a covered loss leads to a business interruption, this coverage can reimburse the contractor for lost income during the downtime.

**Accessories and attachments:** Coverage can be extended to include accessories and attachments that are essential to the operation of the insured equipment.

It's important to note that Contractors Equipment Insurance typically does not cover liabilities resulting from accidents or injuries caused by the equipment.

For such liabilities, contractors usually need General Liability Insurance or Workers' Compensation Insurance.

*Insurance required for contractors: Guarantee insurance to ensure compliance with the contract.*

Guarantee insurance, also known as contract guarantee insurance or performance bond insurance, is a type of insurance designed to ensure compliance with the terms and conditions of a contract.

It provides financial protection to the beneficiary of the contract (usually the project owner or client) if the contractor fails to fulfill their contractual obligations.

When a contractor enters a contract, the project owner may require them to provide a guarantee or bond as a form of security.

This guarantee serves as a promise that the contractor will complete the project as specified in the contract, and it provides reassurance to the project owner that they will be compensated if the contractor fails to meet their obligations.

Here's how guarantee insurance works:

Contractor obtains the insurance: The contractor purchases guarantee insurance from an insurance company or a surety bond provider.

The insurance provider evaluates the contractor's financial stability, track record, and ability to fulfill the contract before issuing the guarantee.

Terms of the guarantee: The guaranteed insurance policy outlines the terms and conditions under which the insurance provider will compensate the project owner if the contractor defaults on the contract.

These terms typically include the amount of coverage, the scope of the guarantee, and the triggering events that would lead to a claim.

Contractor defaults on the contract: If the contractor fails to fulfill their contractual obligations, such as non-completion of the project, substandard work, or financial default, the project owner can make a claim on the guaranteed insurance policy.

Insurance provider compensates the project owner:

If the claim is valid and falls within the terms of the policy, the insurance provider will compensate the project owner up to the coverage amount specified in the policy.

The insurance provider may then seek reimbursement from the contractor for the amount paid out.

Guarantee insurance provides protection to the project owner against financial losses resulting from the contractor's failure to meet their obligations.

It helps ensure that the project owner can complete the project or hire another contractor without incurring significant additional costs. It's important to note that guarantee insurance is different from liability insurance.

Liability insurance protects against claims for injury or property damage caused by the contractor's actions, whereas guarantee insurance focuses on the contractor's performance and compliance with the contract terms.

*Insurance required for contractors: Surety insurance to replace the retention with repair funds.*

Surety insurance is a type of insurance that provides financial protection to the project owner or client if the contractor fails to fulfill their contractual obligations.

It is commonly used in construction projects to replace the traditional practice of withholding retention funds with repair funds.

In many construction contracts, a certain percentage of the contract price is withheld by the project owner as retention funds.

These funds are typically intended to provide a form of security for the project owner in case the contractor does not complete the project satisfactorily or fails to address any defects or issues that may arise during the warranty period.

Surety insurance offers an alternative approach to retention funds by providing a guarantee from a third-party surety company.

Here's how it works:

**Contractor obtains surety insurance:** The contractor purchases surety insurance from a surety company.

The surety company evaluates the contractor's financial stability, track record, and ability to fulfill the contract before issuing the insurance.

**Terms of the surety insurance:** The surety insurance policy outlines the terms and conditions under which the surety company will provide financial compensation to the project owner if the contractor fails to meet their contractual obligations. This includes addressing defects or issues that may arise during the warranty period.

**Replacement of retention funds:** Instead of withholding retention funds from the contractor, the project owner relies on the surety insurance as a form of financial security.

The surety insurance replaces the need for retention funds by providing assurance to the project owner that they will be compensated if the contractor fails to fulfill their obligations.

**Contractor defaults on the contract:** If the contractor fails to meet their obligations, such as non-completion of the project or failure to address defects, the project owner can make a claim on the surety insurance policy.

**Surety company compensates the project owner:** If the claim is valid and falls within the terms of the policy, the surety company will compensate the project owner up to the coverage amount specified in the policy.

The surety company may then seek reimbursement from the contractor for the amount paid out.

By utilizing surety insurance instead of retention funds, the project owner can potentially free up cash flow that would otherwise be tied up in withheld funds.

It also provides an additional layer of financial protection and assurance in case the contractor fails to fulfill their obligations.

It's important to note that surety insurance does not replace other types of insurance coverage, such as liability insurance or professional indemnity insurance, which may still be required to address other types of risks associated with the project.

Insurance required for contractors: Policy of guarantee for financial advance and/or collection.

A policy of guarantee for financial advance and/or collection is a type of insurance coverage that provides protection to businesses or individuals who have advanced funds or made payments to another party.

This insurance policy safeguards the insured party against the risk of non-repayment or non-collection of the advanced funds.

Here's how a policy of guarantee for financial advance and/or collection typically works:

**The insured party requests a guarantee:** The party who has advanced funds or made payments (referred to as the beneficiary) seeks a guarantee from an insurance provider to protect their financial interest.

**Evaluation and underwriting:** The insurance provider evaluates the request and assesses the risk associated with the guarantee. This involves examining the financial standing and creditworthiness of the party to whom the funds have been advanced (referred to as the debtor).

**Issuance of the guaranteed policy**: If the insurance provider determines that the risk is acceptable, they issue a guaranteed policy to the beneficiary.

The policy outlines the terms and conditions under which the insurance provider will provide compensation to the beneficiary.

**Default or non-repayment**: If the debtor defaults on the repayment or fails to fulfill their obligations, the beneficiary can make a claim under the guaranteed policy.

**Compensation by the insurance provider:** If the claim is valid and falls within the terms of the policy, the insurance provider compensates the beneficiary for the financial advance or payment that was not repaid or collected.

The insurance provider may then seek recovery from the debtor for the amount paid out.

A policy of guarantee for financial advance and/or collection helps protect businesses or individuals from potential financial losses resulting from non-repayment or non-collection.

It provides an added layer of security when dealing with transactions involving significant sums of money or when dealing with parties whose creditworthiness may be uncertain.

*Any other insurance required by current labor legislation.*

*What are the additional clauses required by the Client?*

Generally, the Client requests that insurance policies include the following clauses:

Clause waiving subrogation by the insurer against the Client. (Known as the non-repetition clause)

Clauses that prevent the insurer from modifying and/or canceling the insurances without prior notice to the Client.

We should note that the Client regularly monitors the validity of each of the policies required in the contract document.

# ANALYSIS OF FINANCE COSTS

This includes all disbursements in terms of monetary units of interest, commissions and other expenditures resulting from obtaining loans from financial institutions.

Where work goes beyond the economic potential of the contractor, private capital or credit institutions must be used to provide the contractor with the economic means to undertake the work.

This involves the payment of interest or the involvement of a third party in the benefits of the work.

From the above, it is essential to carry out, with the estimate of the working budget, an economic and financial study consisting of an expenditure budget or an investment plan, and a revenue budget or resources that will be available as work progresses, as payment for the work performed.

The result between the two sets of values, investments, or expenditures on the one hand, and revenues on the other, determines the monetary needs of the work.

The main component of the cost of financing is the one resulting from multiplying the capital financed by the time of financing at the corresponding interest rate.

To determine with some degree of accuracy the cost of financing a project, it is necessary to have an investment plan.

## Investment plan

The investment plan, or estimates, are drawn up based on the workplan.

The steps for developing the work plan and related investment plan are outlined below.

## Work plan

The work plan is obtained from the planning of the project or work, which implies quantifying the time and resources that the project will need.

Planning is essential to describe the action plan to be followed and to define the logical sequence of activities or tasks to be developed.

The planning of a work is represented by a diagram, for instance a Gantt, in which the following elements are examined:

 All the tasks to be carried out and the supplies to be supplied.

 The choice of technologies to be implemented for the development of the project.

 Performance times for each activity.

 To establish the execution times for each activity, it is necessary to determine the productivity of the workers and the related equipment, define the times for the provision of supplies, etc.

Defining the sequence or a logic chain in which activities are developed.

Once the above steps are completed, a preliminary work plan can be created based on the milestones identified by the client or the submission document.

Then you quantify the resource use of each activity and add up the cost of those resources.

Once this is done, cash flow needs may be allocated over time, i.e.: formulate an investment scheme.

This enables us to determine the amount of initial capital required and the funds to be invested periodically during construction.

Once the investment plan has been prepared, it is necessary to determine the dates and amounts of income to be received as payment for the progress of the work performed.

The resultant between the two tables of values, investments or expenses and payments or income (payback) determines which is the cash flow and in which period the need for capital is produced.

As can be seen, the financial cost of a project is linked to the method of payment of the purchaser and the payment terms that the contractor obtains with its suppliers and subcontractors, etc.

The Contractor often offers its suppliers and subcontractors the same payment terms agreed with the Client.

All this analysis is done in conjunction with the company's financial department and defines what is charged in the estimate.

An example of financial costing is given at the end of the book.

# CONTINGENCY RESERVE

The owners or project managers of the company are often tasked with analyzing and defining contingency reserves.

To carry out a project, one must be very careful in the analysis of risk management.

## Risks exist and happen in each project

*The risks are grouped as follows:*

*Identified even if the magnitude of the impact is not exactly known.*

In such cases, the impact of the risks is mitigated by the contingency reserve.

*Unidentified.*

Such risks shall be addressed by the management reserve.

As we can see, the contingency reserve and the management reserve are not the same.

The contingency reserve covers identified risks and forms part of the cost base, while the management reserve covers unidentified risks and is part of the budget.

## Calculation of the Contingency Reserve.

The amount of money, or time, can be estimated in several ways, and often only considers negative risks.

Calculations should consider the likelihood of risks occurring, the financial magnitude of their impacts, and the cost of alternatives.

Project leaders and their teams analyze and define the cash or time reserves to be considered to cover these eventualities.

The contingency reserve is included in the cost baseline, i.e. Base cost = project cost estimate + contingency reserve.

> Base cost = Contingency Reserve + Construction Cost Estimates

## How to determine the cost of contingencies

Typically, we add a contingency reserve to a budget where there is some statistical certainty that unpredictable individual costs will be incurred.

The amount of funds or time intervals allotted for each contingency is a value that balances the accepted risk.

On the case of a project, the need for a contingency reserve is based, for example, on the probability of occurrence and the impact of one or more of the following events:

## Work is being done in an area of volatile weather conditions

The bidder shall cover contingencies when the project is affected by the probability of occurrence and impact of:

Extreme temperatures.

Prolonged heavy rainfall.

Possible flooding.

Frequent winds, etc.

## Potential labor conflicts

In some regions, trade unions are highly radicalized, and this situation increases the possibility of strikes and changes in labor regulations.

## The project will operate in a country with poor economic stability.

Some nations have very unstable economic and political situations.

This causes, for example, changes in the market resulting from frequent and unanticipated increases in prices and interest rates.

The Bidder must protect itself from this situation by considering the likelihood of occurrence and the impact of this event.

## Tight timelines to carry out a project and high penalties for non-compliance

If the project to be quoted is:

A short delivery period and severe penalties in case of non-compliance.

The tenderer shall protect himself against possible losses.

## The designs are not entirely defined in the tender documents

This condition creates uncertainty and, as a result, the need to cover this eventuality.

## Long-term projects

In the case of long-term projects to be implemented in inflationary countries, the client regulates the amount of certifications with a polynomial adjustment formula for cost changes.

The above formula must be analyzed by bidders and, where necessary, protected for cost increases not contemplated in the formula.

In conclusion, this part describes what needs to be assessed to quantify contingency reserves.

## The management reserves

The management reserve is the reserve added to the overall project by senior management to cover unidentified or uncertain events.

These risks are not identified as part of the risk management process.

The management reserve is not included in the cost baseline, i.e.:

The project budget = the cost baseline + the management reserve.

> **Project Budget = Base cost + Management Reserve**

The reserve is maintained until the end of the project.

## Example

The supplier of a basic material for the project closes unexpectedly due to economic problems and it is necessary to look for another supplier of that material that meets the requirements of the Technical Specifications of the Bid.

In this case, the project manager must report the even to senior management to use the Management Reserve to solve this unforeseen event.

# PROFIT MARGIN

The profit margin is a financial metric used to assess the profitability of a business or company.
It represents the percentage of profit a company makes from its total revenue after deducting all expenses associated with the production and sale of goods, or services.

## Calculate profit margin

Calculating the profit margin in construction or assembly works involves analyzing the revenues generated from the project and subtracting all the associated costs to arrive at the net profit.

The formula for the calculation of the profit margin is as follows:

In the following figure we highlight the Item Profit margin.

$$\frac{\text{Net Profit}}{\text{Total Revenue}} \times 100 = \text{Profit Margin}$$

## Where:

– Net Profit refers to the total profit earned by the company after deducting all project-related expenses, such as labor costs, materials, equipment, subcontractor fees, permits, overhead costs, and any other project-specific expenses.

– Total Revenue represents the total amount of money received by the construction or assembly company from the client or the project sponsor for the completed work.

It's important to note that the profit margin in these works can vary significantly depending on various factors, including the type of project, size, complexity, geographic location, and competition within the industry.

To cover their operational costs, ensure financial stability, and invest in growth and expansion, construction companies typically aim to maintain a healthy profit margin.
Inefficiency, poor cost management, or intense competition can lead to low profit margins, while effective project management, cost control, and pricing strategies can lead to high profit margins.

Construction and, or assembly companies must be careful in estimating the project costs and pricing their services accurately to maintain a satisfactory profit margin while also remaining competitive in the market.
Monitoring and analyzing profit margins regularly can help these companies make informed decisions to improve their financial performance and overall profitability.

## Exploring Profit Margin: A Tool for Your Business | Example

Suppose a company located in the vicinity of an oil region produces and sells the same type of pre-assembled Skid as the one shown in the following figure of the net page.

This company manufactures, moves, and assembles the Skid on an existing foundation, it has the following financial balance:

Cost of manufacturing, transferring, and assembling equal to $80,000.

Sale Price $100,000

Profit Margin 20%

The company is invited to quote in another oil region, much further away from its factory and with higher assembly costs.

What is the new sale price if you want to maintain a 20% profit margin?

*Financial balance*

The cost of manufacturing, transfer, and assembly is equal to $90,000.

Profit Margin 20%

New selling price = $A.

problem solved using formula.

**Profit Margin = (Net Profit / Total Revenue) x 100**

replacing

20% = (A – $90,000/A) x 100,

if we clear we have that A = $112500

we check.

*PM = ($112,500 – $90,000/$112,500) x 100 = 20%*

Maintaining the PM in this case implies increasing the sale price to $112,500.

## Typical Profit – How to Calculate Profit Margin

Constructing and, or assembly projects typically have a percentage of benefits that is between 5 and 30%.

Commonly, high percentages are common in small and/or risky jobs.

The usual is:

For small and/or difficult jobs, the margin to be charged is between 20 and 30%.

On medium-sized jobs and with less risk, margins of 15 to 20% are common.

From large and low risk works, the margin is 10 to 15%.

In very large and low risk works, the margin is 5 to 10%.

*Each company defines its percentage of profits or earnings according to:*

The stability of the currency in which we quote it.

The profitability, which is the percentage of return on investment measured over time.

Occupation of its operational capacity.

The size and risk of the work.

The intention to achieve continuity of work with that client, etc.

*The profits earned by a company belong to its owners.*

In the execution of a project, each company puts into play his prestige, capacity, experience, capital, equipment, and other relevant factors.

The profits earned by a company belong to its owners, who are the shareholders or stakeholders in the business. After deducting all expenses, including operating costs, taxes, and interest, the remaining amount is the net profit.

This net profit is essentially the company's earnings or income.

In the case of a publicly traded company, the ownership is distributed among shareholders who hold shares of the company's stock.

The proportion of ownership that a shareholder has been determined by the number of shares they hold relative to the total number of outstanding shares.
When a company generates profits, it can do one of several things with the earnings:

Distribute dividends: The company may decide to distribute a portion of its profits as dividends to its shareholders.

Dividends are a way of sharing the company's success and rewarding shareholders for their investment.

Retain earnings: Instead of distributing all profits as dividends, the company might choose to retain a portion of the earnings to reinvest in the business for growth and expansion.
These retained earnings can be used for research and development, capital investments, debt reduction, or other strategic initiatives.

Buyback shares: The company can use its profits to buy back its own shares from the market. This reduces the number of outstanding shares and effectively increases the ownership stake of existing shareholders.

Pay off debt: Profit can be used to pay down existing debts, reducing the company's interest expenses, and improving its financial health.

The decision on how to use the profits depends on the company's financial goals, growth strategy, and the preferences of its management and board of directors.

## Markup

Markup is the amount by which the cost of a budget is increased to obtain the selling price.
Determining the markup value in construction is a crucial aspect of pricing a project and ensuring that a construction company can cover its costs and generate a reasonable profit.

## Markup Formula

$$\frac{\text{Price} - \text{Cost}}{\text{Cost}} = \text{Markup}$$

We apply the markup to the quotation as follows:

## As a percentage of the final cost of the estimate = Cost markup

*Final price (no taxes) = cost + cost * (% cost markup)*

*If the cost is 1 and the markup 20% the price is 1.20.*

## As a percentage of its selling price = Mark-up on price

In this case we will obtain the final price as follows.

We denote the unknown final price as Pf.

*Pf = cost + Pf *(% markup)*

From the formula, we clear the final price (Pf), and we have:

*Pf = cost / (1 – (% markup)*

*If the cost is 1 and the markup is 20%, the final price is 1.25.*

## Criteria to determine the Markup value.

*The margin that a company sets usually depends on:*

From the previous experience that each company has in the execution of similar works.

The characteristics of the work.

Project Size and Duration: The size and duration of the project can also impact the markup.

Longer projects might require a higher markup to account for increased administrative and financing costs over time.

The conditions of payment. If the construction company needs to secure financing or take loans for the project, the interest on the loans may need to be factored into the markup.

Market Conditions: The level of competition in the construction industry and the current market conditions can influence the markup value.

In highly competitive markets, companies might have to lower their markup to win contracts, while in less competitive markets, they may have more flexibility to set higher markups.

The interest or need for the proponent to execute this work.

Company Reputation and Expertise: Established and reputable construction companies with a track record of successful projects and specialized expertise may command higher markups compared to newer or less experienced firms.

**The risk that the bidder perceives as reasonable to assume.** Projects that involve higher levels of risk, complexity, or uncertainty may warrant a higher markup to account for potential contingencies and challenges that could arise during construction.

It's essential for construction companies to strike the right balance to remain competitive while ensuring profitability and sustainability in the long run.

# COST OF TAXES

## Tax for Construction Contractors

When preparing a proposal, it is important to consider the tax burdens that affect the finances of the project.
In this publication, we will give an overview of the types of taxes applicable to the construction sector.

The tax obligations for construction contractors can vary depending on several factors, including the contractor's legal structure (sole proprietorship, partnership, limited liability company, or corporation) and the country or region in which they operate.

Cost per tax is the last step to be completed to successfully determine the bid price.

## What is a Tax?

A tax is a compulsory financial charge or levy imposed by a government on individuals, businesses, or other entities to fund public expenditures and government programs.

Taxes are a primary source of revenue for governments at various levels (local, regional, national) and are used to finance public services such as infrastructure development, healthcare, education, defense, social welfare programs, and more.

Taxes are typically enforced through legislation and collected by tax authorities or government agencies responsible for revenue collection.
The specific tax laws and regulations can vary from country to country, and even within different jurisdictions within a country.

## Taxes can be imposed on different types of activities and income, including:

Personal Income Tax: A tax on the income earned by individuals, which is typically calculated based on a progressive tax rate system, where higher-income individuals are subject to higher tax rates.

Corporate Income Tax: A tax on the profits earned by businesses or corporations. Corporate tax rates can vary depending on the jurisdiction and the size of the company.

Sales Tax or Value Added Tax (VAT): A tax imposed on the sale of goods and services. It is usually a percentage of the purchase price and is collected by businesses at the point of sale.

Property Tax: A tax levied on the value of real estate or other properties owned by individuals or businesses.

Property taxes are usually assessed by local governments and used to fund local services like schools, roads, and public facilities.

Payroll Tax: A tax withheld from employees' wages and paid by employers to fund social security programs, healthcare, and other benefits.

Excise Tax: A tax imposed on specific goods or activities, such as tobacco, alcohol, fuel, luxury items, or environmentally harmful products.

Excise taxes are often used to discourage certain behaviors or to fund specific programs or initiatives.

These are just a few examples of the various types of taxes that exist. Taxation is a complex subject, and governments may use different tax structures and policies to achieve their fiscal objectives.

It's important to understand and comply with the tax laws and regulations of your specific jurisdiction to avoid penalties and ensure proper contribution to public finances.

*In economic terms, taxes, transfer wealth from individuals or businesses to the government.*

## Tax burdens to be considered in the proposal

In general, the Client is not liable for any tax, fee, or contribution, whether national, provincial, municipal, or foreign, levied on the contractor during performing the contract.

This is a common practice in all contracts, so the bidder must have adequate advice on the type and cost of taxes that apply.

Sometimes the contracting party is also a withholding agent, so it will deduct from each payment the amount that corresponds according to the legal regulations in force for Income Tax, Gross Income Tax and VAT.

It's crucial to note that tax obligations can be complex, and the above information provides only a general overview.

Construction contractors should consult with a tax professional or accountant to ensure compliance with all applicable tax laws and regulations.

# EXAMPLE OF A DIRECT ESTIMATE OF PERSON-HOURS

Example of direct labor. Calculation of labor needed to perform a reinforced concrete base.

In this example, we calculate the time required to build a reinforced concrete foundation.

Here, the concrete is provided by a semi-automatic concrete facility located on the site.

The foundation to be constructed has the following dimensions:

Height 0.50 meters.

Length 3 meters.

Width 2 meters.

This means that the foundation has a volume of 3 m3, with a quantity of 100 kg of iron per m3 of concrete.

The next figure illustrates the filling process of the structure already in place with the formwork and reinforcement.

This calculation excludes the time required for excavation, clearing and backfilling of the soil.

A table summarizing the hours spent.

*Example of direct labor. The chart shows the number of hours of work consumed to build the base.*

| Estimation of the man-hours required to build a 2*3*0.50 m high reinforced concrete base ||||| 
|---|---|---|---|---|
| items | Quantity | Unit | Performance | Mhr |
| Construction iron amount 100kg/m3 | 300 | kg | 10 kg/Mhr | 30.00 |
| Elaborated concrete to be poured in the base | 3 | m3 | 1.50 Mhr/m3 | 4.50 |
| Formwork manufacture, to be used only once | 5 (Wet wall) | m2 | 0.35 m2/Mhr | 14.28 |
| Total | | | | 48.78 |

*Hours of work, according to the chart to do this base: 30 Mhr + 4.5 Mhr + 14.28 Mhr = 48.78 Mhr*

We must adjust this value with the variable times specific to each task.

## Variable Times

Labor used to transport equipment and materials to the base site.

Labor for layout and control.

Practice layer depression time if applicable.

Work to excavate and subsequently fill soil.

Work on concrete cleaning.

Work for curing, disassembling, and cleaning.

Added work if we construct the base in an operating plant.

In this case, we consider the site location to be isolated and the performance of the work to be independent of another operating installation.

Time to be added for meteorological conditions that differ from normal conditions.

In the standard state, we perform tasks in a moderate climate (i.e., with average temperatures between 5°C and 40°C) and winds less than 30 km/hour.

# EXAMPLE OF CALCULATING THE COST OF EQUIPMENT PER HOUR

Example of an hourly cost calculation for a pickup truck.

In this example, one will calculate the hourly cost of a Ford F150 4x4 6-cylinder truck.

Note that the reader must substitute the values used in this example to adapt them to those in effect in his area.

Fixed cost formula

Net Capital Value = $V_n$.

$V_n$ = Net value = $V_{ad} - P_4 - P_e$

Where:

$V_{ad}$ is the acquired value.

$P_4$ is the cost to replace four tires annually.

$P_e$ is the value of special pieces and accessories.

The table below gives the parameters used for the calculation.

| ITEM | INFORMATION |
|---|---|
| Acquistion value | Vad = USD 23000 |
| Economical or usefur life | Vu = 6000 hours |
| Value of tires | P4 = USD 1000 |
| Special parts value | Pe = USD 0 |
| Net value of the equipment | Vn = Vad-P4-Pe = USD 22000 |
| Recovery Factor | R = 20% |
| Resale or redemption value | Vn*r = USD 4400 |
| Annual working hours | 2000 hours |

# Determining the fixed cost of the equipment

Formulations to be used.

*Resale value of machinery and vehicles.*

Vrev = The resale value of a vehicle at the end of the useful life.

In the case of the F150, we deducted a resale value of 20% of the acquisition value.

*Depreciation*

The value to be allocated to amortization is the result of the following formula:

D= [Vn-Vrev/Vu]

*Medium interest on investment = Im.*

Mean investment interest at an annual interest rate = i.

The burden to be added to the cost of the equipment for the average interest of the investment shall be:

The average equipment value of annual interest, divided into 2000 hours per year.

Im = {[Vn + Vrev]/2 * 2000} *i

*Cost of insurance, taxes, and warehousing.*

Insurance, taxes, and annual storage fees are:

Cost of insurance: 5.5% of the average value of the equipment.

Cost for taxes: 1.5 per cent of the average value of the equipment.

The cost to store 1.5% to 2% of the average value of the equipment.

Hourly costs are calculated by dividing the amounts by the number of annual hours of work, 2000 hours.

*Maintenance and repair of machines.*

*Maintenance for repairs and spare parts.*

We attribute the hourly maintenance cost of repairs and replacement parts for the F150 pickup to a value equal to 50% of its depreciation value.

The formulas used to calculate the fixed cost per hour of the F150 are presented in the follow table.

| FIXED HOURLY COSTS | | |
|---|---|---|
| ITEM | FORMULA | PARTIAL COST |
| Depreciation value | D = [ Vn-Vrev] / Vu | [22000-4400] /6000 = USD 2.93 |
| Interest on the investment, rate 3% per year | Im = [(Vn+Vrev)/2*2000]*i | [(22000+4400)/2*2000]*3% = USD 0.20 |
| Comprehensive insurance, rate 5.50% of average equipment value | Sm = [(Vn+Vrev)/2*2000]*s | [(22000+4400)/2*2000]*5,5% = USD 0.36 |
| Cost for taxes. Annual rate 1.5% of average equiment value | Taxes = [(Vn+Vrev)/2*2000]*1.5% | UDS 0.10 |
| Cost for storage. Annual rate 1,50% of average equipment value | Storage = [(Vn+Vrev)/2*2000]*1.5% | USD 0.10 |
| Cost of maintenance and repairs for light work 60% | D*0.60 | 2.93*0.60 = USD 1.76 |
| Total fixed cost per hour | | 2.93+0.20+0.36+0.10+0.10+1.76 = 5.45 |

The fixed cost per hour of the F150 truck in the example is 5.45 USD/hour.

## Hourly Cost per Consumption

*Cost per hour for fuel and lubricant.*

Assumptions:

The truck operates 2,000 hours per year.

The fuel consumption is 8.5 liters/hour.

The value of a liter of fuel is 1 US dollar.

Notwithstanding the above, the main factors influencing fuel consumption are discussed below.

Styles of driving.

Accelerating the vehicle.

Driving speed.

The vehicle's age.

The operational condition of the vehicle.

Ambient temperature

Wind and its direction.

Traffic

Depending on whether the vehicle is driven into the city or on the highway.

The engine rpm.

Load transported by vehicle.

Type & quality of fuel.

Road conditions.

Use of vehicle air conditioner, etc.

Generally, each company records the fuel consumption of each machine, for example, for a month, and then distributes this consumption in the monthly working time of the machine, thus obtaining the hourly fuel consumption of the equipment.

*Lubricant Cost*

From the pickup vehicle, the cost of lubricant consumption is taken as equal to 7% of the hourly cost of fuel.

The correct method for finding out the hourly consumption of oil in a machine is to record the consumption of all lubricants, i.e., engine oil, hydraulic oil, and transmission oil, for changes and refills at the end of each month, and then divide each data into the monthly working time of the machine.

It is common to associate lubricant consumption with fuel consumption over the same period and to express lubricant consumption as a percentage of fuel cost per hour.

*Tires Cost*

In this example, we consider that the truck tires have an exceptionally low lifespan of only one year because the vehicle runs on bad roads.

We take the cost of the four truck tires per year to be: $P_4 = \$1,000$

*Tire recommendations*

In general, tire rolling resistance accounts for about 15 percent of total fuel consumption.

When faced with tires that are similar in terms of grip and ride comfort, it makes sense to use tires that have lower rolling resistance because they provide greater fuel efficiency.

In all tires, lower pressure means greater rolling resistance and higher fuel consumption.

*The table below summarizes the cost per hour of consumption of the Ford F150 4 4 from our example.*

| HOURLY CONSUMPTION COST | |
|---|---|
| Fuel consumption | 8.5 Liters/hour |
| Fuel price | USD 1 * liter |
| Hourly fuel consumption cost | 8.50 * 1 = UDS 8.50 |
| Lubricant costs | 7% * 8.5 = USD 0.60 |
| Cost per tires | P4/2000 = 1000/2000 = USD 0.50 |
| Total | 8.50 + 0.60 + 0.50 = USD 9.60 |

# Cost per hour for a pick-up truck – Operating cost

Cost per hour for operator.

*Labor costs*

The cost of the hourly wage of the worker, plus the social security contributions multiplied by the assigned percentage, is equal to the hourly cost of the worker.

The remuneration and social security contributions of operators vary considerably according to the country or region in which the work is carried out.

In this example, we will take a salary of 10 USD and social expenses of 65%.

Therefore, the operator's remuneration and social security contributions are USD 16.50 per hour.

*Cost of Surveillance.*

The cost per hour of surveillance is equal to 10% of the labor cost.

As such, the cost per monitoring is: 16.50*10% = 1.65 USD

The cost of surveillance applies particularly to high-value machinery.

*Summary of total cost per hour*

The total hourly cost of the Ford F150 4×4 6-cylinder pickup truck in the example is:

Fixed cost per hour = USD 5.45.

Cost of consumption per hour = $9.60. Labor cost per hour = USD 16.50.

Surveillance Labor Costs = $1.65.

*Total, hourly cost = 5.45 + 9.60 + 16.50 + 1.65 = US$ 33.20.*

# FINANCIAL COST EXAMPLE

Here, we will analyze the financial requirements to complete the work described below.

This is a small civil structure, including the construction of two reinforced concrete bases of 10 m3 each.

To perform this analysis, we develop a Gantt chart showing each of the tasks to be performed during the construction period.

Then we load into the Gantt the flow of funds we need to execute the job.

*In summary, we put together a rough investment plan.*

Once the investment plan is in place, it is necessary to determine the dates and amounts of revenues to be received while the work is being performed as payment for the work performed.

*The result between the two tables of values, investments or expenditures, and payments or income (reimbursement) determines the capital requirement.*

Therefore, this provides us with the liquidity of the project.

## INVESTMENT PLAN

| Investment Plan | % | Month 1 | Month 2 | Month 3 | Month 4 | Month 5 | Month 6 |
|---|---|---|---|---|---|---|---|
| Assembly of the workshop and mobilization | 8% | | | | | | |
| Purchase of construction iron and the formwork wood | 8% | | | | | | |
| Excavation for two bases of 10 M3 each | 8% | | | | | | |
| Reinforcement iron and formwork assembly | 16% | | | | | | |
| Purchase and casting of elaborated concrete | 46% | | | | | | |
| Demoulding and filling with soil in the area of the base | 6% | | | | | | |
| Dismanteling of the workshop and demobilization | 8% | | | | | | |
| **Summary:** | | | | | | | |
| Expenses per month (these expenses include overhead) | | 22% | 64% | 6% | | | |
| Accumulated costs | | | 86% | 92% | 100% | | |
| Expected income from payments: Monthly certification of work in progress, payable 30 days f/f (includes 30% of profits) | | | | 22%*1.3 = 28.6% | 64%*1.3 = 83.2% | 6%*1.3 = 7.8% | 8%*1.3 = 10,4% |
| Accumulated earnings | | | | 28.6% | 111.8% | 119.6% | 130% |

# Cash flow of the project

The chart above shows the following:

*The monthly expenditure necessary to carry out the work.*

*Income receivable:*

We assume that we receive the payment 30 days after the month-end certification date.

In the interest of simplicity, we consider that the profit is constant and exceeds the expenditure by 30%.

This example shows that the contractor must have his own funds for months one, two and three.

If he does not have his own funds, he must finance them to be able to do the work.

## Cash flow

For clarity, we summarize the data in a chart showing the flow of funds needed to build the work.

| Month | Expenses expressed as a percentage | Income expressed as a percentage | Accumulated balances expressed as a percentage |
|---|---|---|---|
| 1 | 22 | 0 | -22.00 |
| 2 | 64 | 0 | -86.00 |
| 3 | 6 | 28.60 | -63.40 |
| 4 | 8 | 83.20 | 11.80 |
| 5 |  | 7.80 | 19.60 |
| 6 |  | 10.40 | 30.00 |

The need for money for the first three months of work generates financial costs.

*Further Reading:*

## Estimator's Piping Man-hours Tool

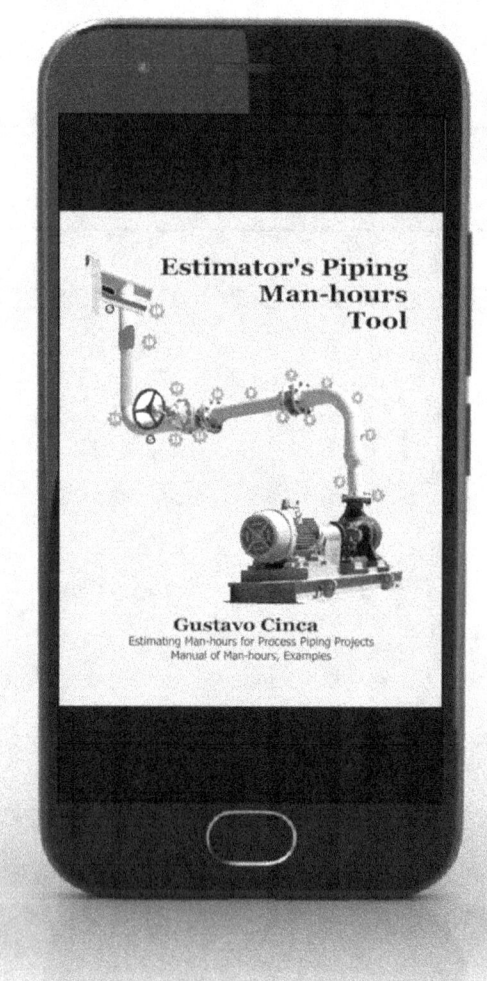

In this book, the author details a quick and easy method to calculate the man-hours required to assemble the carbon steel process piping. The records in the tables reproduce the yields used by the author during his professional career.

**NEW!**

# Bolted Flange Joint

'uds and Gaskets: Recommended Practices for the Assembly of a Bolted Flange Joint

*In this publication you will find a large number of pictures, basic principles and descriptions, which will help you to solve the most common questions that arise during the process of assembling a flanged joint.*

*The book is intended as an orientation tool for novices and an easy-to-read tool for experienced employees.*

# Types of Valves in Piping

Types of Valves - Tables to Estimate Man-hours of Assembly

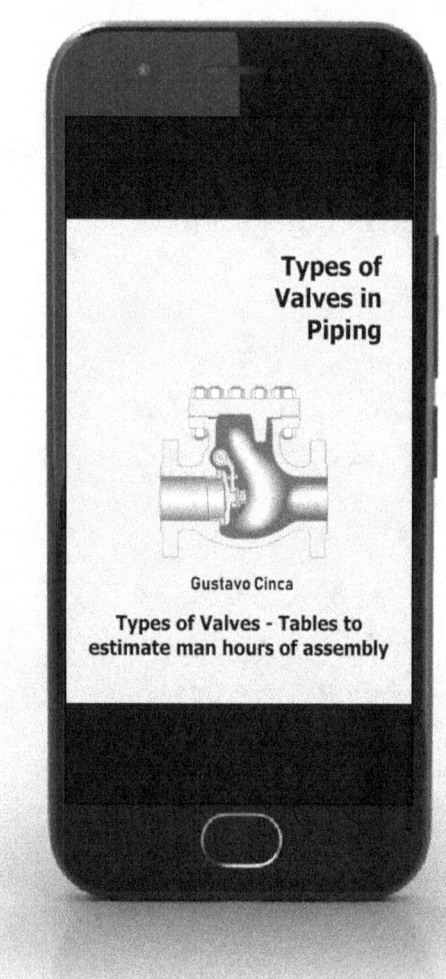

*Suitable for beginners. This publication outlines traditional valves used in pipeline systems. The manual includes, as a supplement, tables with records of man-hours required to assemble flanged, flanged, butt-welded, and wafer*

# About the Author

Gustavo Miguel Cinca was born in San Rafael (Argentina) and obtained his diploma with distinction in chemical engineering.

During his professional career, he worked as a construction manager and, finally, he created *and presided over an industrial construction and assembly company for more than 20 years.*

Throughout his professional career, he has built chemical processing plants in refineries, pipelines, compressor plants and thermal power plants at home and abroad.
Based on this experience, the author provides the reader with this book.

Read more at Gustavo Cinca's site.

www.ingramcontent.com/pod-product-compliance
Lightning Source LLC
Chambersburg PA
CBHW080501220526
45465CB00006B/2342